ACID-BASE, FLUIDS, AND ELECTROLYTES MADE RIDICULOUSLY SIMPLE

Richard A. Preston, M.D., M.B.A.
Associate Professor of Clinical Medicine and Nephrology
Director, Division of Clinical Pharmacology and
Pharmacokinetics and Clinical Research Center
Department of Medicine
University of Miami School of Medicine

MedMaster, Inc., Miami

Published by
MedMaster, Inc.
P.O. Box 640028
Miami, FL 33164

ISBN # 0-940780-31-3
                        Cover art by Conrad Barski

CONTENTS

Dedication:

To Ruth, Dick, Scott, Joanne, Rianna, Glenn, Beth, Luke, Grace, Mercy, and Peace Preston

Acknowledgments:

I would like to thank the following colleagues for their helpful comments and suggestions:

Nicholas P. Bell, Medical Student
Jonathan B. Berger, M.D.
Bernhard L. Brijbag, D.O.
Juan Pablo DeTorres, M.D.
Bruno S. Fang, M.D.
Stephen Goldberg, M.D.
Juan A. Guardia, M.D.
Carlos Javier Guzman, M.D.
Jamey Howitt, Medical Student
Michelle Lister, M.D.
Neil Love, M.D.
Barry J. Materson, M.D.
Richard A. McReynolds, M.D.
Sandra L. Mostaccio, M.D.
Nicolas Musi, M.D.
Olayemi Osiyemi, M.D.
James R. Oster, M.D.
Denise Pereira, M.D.
Guido O. Perez, M.D.
Glenn G. Preston, M.D.
Irwin Singer, M.D.
MaryAlice Yoham, M.S.N., A.R.N.P.

PREFACE

Acid-base, Fluids, and Electrolytes Made Ridiculously Simple is intended to be a brief, easy to read book that provides the clinician with a straightforward approach to solving even the most complex fluid, electrolyte, and acid-base problems. It should be useful to medical students, interns and residents, nurses, nephrology fellows, primary care physicians, surgeons and other clinicians responsible for IV fluid therapy.

Handling acid-base, fluid, and electrolyte problems is a common and essential part of the everyday practice of medicine, but I am amazed at how much difficulty these problems present to clinicians. As a student and house officer, I read a number of very good texts on the subject of acid-base, fluids, and electrolytes, but when faced with an actual clinical problem, I often had trouble implementing what I had read. I did not have a consistent, simple, and easy to apply approach to the diagnosis and treatment of acid-base, fluid, and electrolyte problems. Just as it is important to develop a routine for performing the physical examination or for reviewing an ECG, it is essential to have a systematic way of approaching acid-base and electrolyte problems. This approach should be based upon a firm grasp of pathophysiology, but should not be bogged down by excessive scientific detail.

That is why I wrote this book.

More than half of this book consists of clinical examples and exercises. The text of each chapter contains a brief discussion of the key elements of diagnosis and treatment of a specific electrolyte or acid-base disorder. The numerous practice exercises at the end of each chapter, in addition to expanding on concepts developed in the text, help the reader learn a systematic step-by-step approach to solving even the most difficult cases.

This book assumes a minimum requisite background in renal physiology. A rigorously trained renal physiologist might have a few extrasystoles while reading this book: I have "rounded off" a few corners of pathophysiology that I felt tended to get in the way of handling clinical problems. Many of the methods in this book for managing electrolyte and acid-base disorders are based upon approximations. There are often *many* mechanisms involved in producing a particular electrolyte or acid-base disorder, but I usually mention only one or two of the most important and easiest to remember. *Remembering all*

1

the mechanisms involved won't help me if I don't know how to approach a case. My feeling is that it is better to know how to approach and solve clinical problems, rather than to get stuck on the fine points of pathophysiology.

I am grateful to Laurence B. Gardner, M.D., for his superb didactic lectures and bedside teaching on clinical electrolyte and acid-base physiology. I am indebted to Barry J. Materson, M.D, and Guido O. Perez, M.D., for their careful reviews of the manuscript and for their helpful comments and suggestions. My special thanks to a true scholar of fluids and electrolytes, James R. Oster, M.D., for his meticulous and painstaking page-by-page review of the book and for his valuable additions and corrections. Finally, I thank Stephen Goldberg, M.D., for his patience and guidance, and for teaching me so much about how to "keep it simple."

CHAPTER 1. THE BASICS

This chapter briefly reviews the physiology that is key to understanding clinical water, electrolyte, and acid-base disorders. This review lays the groundwork for a more complete understanding of the pathophysiology, diagnosis, and treatment of disorders that are discussed in later chapters. A brief summary of renal tubular physiology is presented in **Fig. 1-1.**

The Body Fluid Compartments

Total body water (TBW) makes up about 60% of body weight in men and about 50% in women (see **Fig. 1-2**). These percentages decline with aging, as the percentage of body fat increases. Approximately 2/3 of total body water is located in the intracellular compartment and constitutes the **intracellular fluid volume (ICFV)**. About 1/3 of TBW is located in the extracellular compartment and comprises the **extracellular fluid volume (ECFV)**. The interstitial fluid volume comprises about 3/4 of the ECFV, and the plasma volume is about 1/4 of the ECFV. The plasma volume is maintained to a large extent by the oncotic effects of plasma proteins. Water passes freely and rapidly between all these compartments in response to changes in solute concentrations to maintain osmotic equilibrium between the compartments. Therefore, the osmolalities in all compartments are equal.

Sodium and Water Pathophysiology

The following approach to sodium and water pathophysiology may be different from what you have studied in the past. I believe it is very important to assess sodium status (which determines the volume of the extracellular fluid compartment) and water status (which determines the serum sodium concentration) separately. Thus it is very important to ask in each case: "does this patient have 1) a problem with sodium, 2) a problem with water, or 3) problems with *both* sodium *and* water?" This is a very important concept that will be developed in this chapter and used to solve complex problems later in the book. The methods for solving electrolyte problems presented in this book will work *consistently* for you once you have mastered them. Just follow along and be sure to do the examples at the end of the chapters.

3

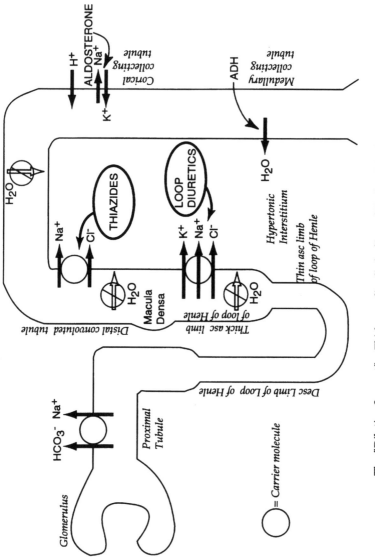

The "Diluting Segment" = Thick ascending limb of loop of Henle + Distal convoluted tubule

FIGURE 1-1.

FIGURE 1-2.   The Body Fluid Compartments

ECFV (1/3 TBW)	ICFV (2/3 TBW)
Sodium 135–145 mEq/L Potassium 3.5–5.0 mEq/L Chloride 95–105 mEq/L Bicarbonate 22–26 mEq/L Glucose 90–120 mg/dl Calcium 8.5–10.0 mg/dl Magnesium 1.4–2.1 mEq/L Urea nitrogen 10–20 mg/dl	Sodium 10–20 mEq/L Potassium 130–140 mEq/L Magnesium 20–30 mEq/L Urea nitrogen 10–20 mg/dl

Women: Total body water (TBW) = .5 × Body weight (kg)

Men: Total body water (TBW) = .6 × Body weight (kg)

Intracellular fluid volume = 2/3 TBW

Extracellular fluid volume = 1/3 TBW

Calculated osmolality = 2 × [sodium] + [glucose]/18 + [Blood Urea Nitrogen]/2.8

Osmolal gap = $OSM_{(measured)}$ − $OSM_{(calculated)}$

The systems that regulate the amounts of sodium and water in the body act together to

- keep the *concentration* of extracellular sodium within a narrow range (135–145 mEq/L)
- keep the volume of the extracellular fluid compartment (ECFV) within reasonable limits.

Sodium Physiology—Regulation of the ECFV

Sodium is the major extracellular cation and is responsible for most of the osmotic driving force that maintains the size of the ECFV.

The total amount of sodium in extracellular fluid (ECF) is the major determinant of the size of the ECFV.

If the total amount of sodium in the ECF increases, so will the size of the ECFV, and ECFV overload will eventually result. The edematous states—congestive heart failure, cirrhosis of the liver, and nephrotic syndrome—are examples of disease states with increased amounts of sodium in the extracellular fluid compartment causing ECFV overload (also termed simply "volume overload"). The increased amount of ECF sodium leads to expansion of the ECFV and the expanded ECFV presents clinically as edema. Other clinical indicators of ECFV overload are pleural effusions, pulmonary edema, and ascites.

If the total amount of sodium in the extracellular fluid compartment decreases, so will the size of the ECFV, and ECFV depletion will eventually result. ECFV depletion (also termed simply "volume depletion") is manifested by poor skin turgor, tachycardia, and an orthostatic fall in blood pressure. *ECFV overload results from too much sodium in the ECF compartment, and ECFV depletion results from too little sodium in the ECF compartment.*

Because sodium is largely confined to the ECF compartment, the amount of sodium in the extracellular fluid compartment is sometimes referred to as **total body sodium.** This term is an approximation because there is a relatively small amount of sodium in the intracellular space. If total body sodium increases, the ECFV will increase and eventually edema will develop. If total body sodium decreases, the ECFV will decrease and eventually ECFV depletion will develop.

The balance between sodium intake and sodium excretion by the kidney determines the amount of sodium in the ECF compartment and therefore the size of the ECFV. The kidney normally adjusts sodium excretion to keep the size of the ECFV within an acceptable range. When the ECFV increases, the kidney increases sodium excretion to prevent ECFV overload. When ECFV decreases, the kidney decreases sodium excretion to prevent ECFV depletion.

Three main systems regulate total body sodium and therefore the size of the ECFV. Each of the three systems has an afferent (sensory) and efferent (effector) limb of sodium control. The afferent limb senses the size of the ECFV and the efferent limb increases or decreases renal sodium excretion accordingly.

- Receptors located in the juxtaglomerular cells of the kidney sense changes in renal perfusion and respond by producing changes in the release of renin, thereby activating the **renin-angiotensin-aldosterone system.** Renin is released in response to decreased renal perfusion. Renin then acts to convert angiotensinogen to angiotensin I, which is converted to angiotensin II by angiotensin-converting enzyme. Angiotensin II directly promotes sodium retention by the kidney and causes release of aldosterone by the zona glomerulosa of the adrenal cortex. Aldosterone promotes sodium retention by the distal nephron.
- Volume receptors are located in the great veins and in the atria and are sensitive to small changes in venous and atrial filling. Activation of volume receptors by increased atrial filling results in release of atrial natriuretic factor, which promotes renal sodium excretion.
- Pressure receptors are located in the aorta and in the carotid sinus. ECFV depletion stimulates these receptors, which activate the sympathetic nervous system and lead to renal retention of sodium.

It is not essential to memorize the details of the pathways linking changes in the size of the ECFV (which is determined by total ECFV sodium) to changes in sodium excretion by the kidney. *The important concept is that normally when the ECFV increases, mechanisms to increase sodium excretion are activated to prevent ECFV overload; and when ECFV decreases, pathways are activated that promote sodium retention by the kidney to prevent ECFV depletion.*

The preceding discussion of sodium regulation does not mention the ECF sodium *concentration*. The ECF sodium concentration is determined by the amount of water relative to sodium in the ECF. The serum sodium concentration, which is measured in clinical practice, does not reliably tell us anything about the *total amount* of sodium in the extracellular fluid compartment or the size of the ECFV. The serum sodium concentration only tells us about the amount of water relative to the amount of sodium.

Osmolality and Tonicity

The main solutes of the ECFV are sodium, glucose, and urea. **Serum osmolality** may be calculated approximately from the formula

$$OSM_{(calc)} = 2 \times [\text{sodium concentration}] + [\text{glucose concentration}]/18 \\ + [\text{Blood Urea Nitrogen}]/2.8$$

where the serum sodium concentration is in mEq/L, and the glucose and blood urea nitrogen (BUN) concentrations are in mg/dl. The amount of water relative to sodium in the ECF determines the ECF sodium concentration. Quantitatively, the sodium concentration is by far the major contributor to the total serum osmolality. *Abnormalities in the sodium concentration tell us that there are abnormalities in the regulation of the amount of water in the ECF compartment.*

It is important to understand the difference between osmolality and tonicity. **Osmolality** is determined by the total solute concentration in a fluid compartment. **Tonicity** refers to the ability of the combined effect of all of the solutes to generate an osmotic driving force that causes water movement from one compartment to another. To increase ECF tonicity, a solute must be confined to the extracellular fluid compartment. That is, the solute must be unable to cross from the extracellular compartment into the intracellular compartment, thereby increasing the osmotic pressure and translocating water into the extracellular compartment. Water moves from the intracellular compartment into the extracellular compartment to establish osmotic equilibrium. Solutes capable of causing such movement of water include sodium, glucose, mannitol, and sorbitol, and are thus said to be "effective osmoles." Sodium remains for the most part in the extracellular space because it is pumped out of cells by sodium-potassium ATPase, so that addition of sodium to extracellular fluid causes water to move out of cells and results in cell shrinkage. Therefore, sodium is an effective osmole because it is capable of effecting water movement.

The extracellular sodium concentration is the main determinant of plasma tonicity. Therefore, when tonicity increases, it is generally because the extracellular sodium concentration has increased. Hypertonicity is the main stimulus for thirst and **antidiuretic hormone (ADH)** release, important factors in the regulation of total body water. If the sodium concentration rises, thirst (leading to water intake) and ADH release (leading to water retention by the

7

kidney) are stimulated. The elevated sodium concentration tells us that there is too little water relative to sodium.

Glucose is an effective osmole but is normally assimilated into cells. Therefore, glucose does not make a large contribution to serum osmolality or tonicity under normal circumstances. In uncontrolled diabetes mellitus, however, a severely elevated plasma glucose concentration can lead to substantial hypertonicity and water movement into the ECF.

Urea contributes to osmolality, but it easily crosses cell membranes and therefore distributes evenly throughout total body water. Because urea crosses freely from one compartment to another down its own concentration gradient, it does not act to translocate water. Therefore, urea does not contribute to tonicity and is not an effective osmole. Urea will increase the measured serum osmolality, but it crosses cell membranes easily and does not contribute to water movement or cell shrinkage.

Control of tonicity determines the normal state of cellular hydration and therefore cell size. Brain cells are of particular concern. Most of the important symptoms and signs of abnormal tonicity are due to brain swelling in response to hyponatremia or shrinking in response to hypernatremia. If an abrupt decrease in the tonicity of the ECFV occurs, water will move into the intracellular compartment, resulting in cell swelling. Conversely, a rapid increase in ECFV tonicity leads to water leaving the brain cells and brain shrinkage.

The Osmolal Gap

The difference between the measured and the calculated osmolality is termed the **osmolal gap:**

$$OSM\ GAP = OSM_{(meas)} - OSM_{(calc)}$$

Values of greater than 10 mOsm/L are abnormal and suggest the presence of an exogenous substance. A significant increase in the osmolal gap can be helpful as a clue to the presence of a variety of exogenous compounds that do not enter into the calculation of osmolality but are measured as osmotically active by the lab. Knowledge of the osmolal gap can be useful in the emergency room setting as a screen for a variety of compounds in suspected ingestions. Sodium, glucose, and urea do not increase the osmolal gap because they affect both the calculated and the measured osmolalities.

Water Physiology—Regulation of the Serum Sodium Concentration (Tonicity)

ECF tonicity is largely determined by the ECF sodium concentration. The homeostatic controls that add or remove water from the body respond to changes in tonicity of the extracellular fluid compartment and keep plasma tonicity constant. This has the important effect of keeping cellular hydration and size constant. Adequate water intake is a function of an intact thirst mechanism and water availability. A rise in ECF tonicity, usually due to an increase

8

in ECF sodium concentration, leads to the sensation of thirst. This control system is very precise: an increase in extracellular fluid osmolality of only a few mOsm/L will lead to significant thirst. Other stimuli cause thirst: Increased angiotensin II and significant ECFV depletion will produce thirst. Thirst is so powerful a stimulus that it is uncommon for a person with a normal thirst mechanism and access to water to develop hypernatremia.

Renal Water Regulation

The kidney responds to changes in extracellular fluid tonicity by adjusting water excretion. In states of increasing tonicity, the kidney decreases water excretion. The kidney decreases water excretion by producing urine that is concentrated relative to plasma. A defect in urine concentration can lead to an inability to conserve water appropriately and can result in water loss and hypernatremia.

In states of decreasing tonicity, the kidney responds by increasing water excretion. The kidney increases water excretion by producing urine that is dilute relative to plasma. A defect in urine dilution can lead to an inability to excrete excess water and can result in water retention and hyponatremia.

In order for the kidney to regulate water excretion to keep the tonicity (sodium concentration) of the ECFV constant, there must be:

• An adequate glomerular filtration rate (GFR)
• Adequate delivery of glomerular filtrate to the concentrating and diluting segments of the loop of Henle and distal nephron
• Intact tubular concentrating and diluting mechanisms
• Appropriate turning on and off of ADH
• ADH responsiveness of the kidney

Virtually *all* of the clinical disorders producing hypernatremia and hyponatremia may be understood and remembered based on abnormalities of these few mechanisms of water regulation.

Glomerular filtration rate (GFR)

Both urine concentration (leading to conservation of water) and urine dilution (leading to enhanced excretion of water) depend upon an adequate GFR. Simply put, if water and solutes are not filtered to enter the renal tubule, then how can the kidney concentrate or dilute the urine to regulate water balance? A GFR reduced to 20% of normal is roughly where the kidney begins to have trouble with *both* concentration and dilution functions.

Water delivery to the diluting segments of the loop of Henle and distal nephron

If a large proportion of glomerular filtrate is reabsorbed proximally, then sufficient water cannot reach the distal nephron to be excreted. Increased

9

proximal reabsorption of glomerular filtrate can lead to water retention and consequent hyponatremia. Two important situations cause increased proximal reabsorption of water and are important causes of hyponatremia:

• Volume depletion (often from vomiting with continued ingestion of water), leading to increased proximal reabsorption of water
• Edematous states: congestive heart failure, cirrhosis, and nephrotic syndrome in which there is increased proximal reabsorption of water

Renal concentrating mechanism

In addition to reabsorbing 20–30% of the filtered sodium, the ascending limb of the loop of Henle generates the hypertonic medullary interstitium and the medullary concentration gradient that is necessary for the *concentration* of urine. Sodium pumped from the loop of Henle by a sodium-potassium-2 chloride cotransporter into the medullary interstitium provides the osmoles necessary for the hypertonic medullary concentration gradient (**Fig. 1-1**). The hypertonic medullary concentration gradient is necessary for reabsorption of water from the collecting tubule and is therefore necessary for the appropriate concentration of urine. Under the influence of ADH, the collecting tubule is rendered permeable to water. As the tubular fluid passes through the collecting tubule, water leaves the tubule and enters the hypertonic interstitium down its concentration gradient and is reabsorbed. The result is a concentrated urine.

Loop diuretics block the reabsorption of sodium in the loop of Henle and impair the formation of the medullary concentration gradient. Therefore, loop diuretics reduce the ability of the kidney to concentrate the urine. Certain chronic renal diseases also cause renal concentrating defects.

Renal diluting mechanism

Both the cortical segment of the ascending limb of the loop of Henle and the distal tubule transport sodium from the tubular lumen, leaving water behind, because the tubular epithelium is impermeable to water. The net effect is that of pumping sodium out while water remains behind, which makes the tubular fluid more dilute.

In the loop of Henle, sodium, chloride, and potassium are transported out of the lumen by a sodium-potassium-2 chloride cotransporter while water stays behind. This transporter is blocked by loop diuretics.

In the distal tubule, sodium and chloride are transported out of the lumen by a sodium-chloride cotransporter which is very important in producing a dilute urine. This transporter is blocked by thiazide diuretics.

ADH

The presence or absence of ADH is the most important factor determining whether the final urine is concentrated or dilute. ADH is released in response to slight increases in tonicity of the ECFV. Because the sodium concentration is the main determinant of tonicity, changes in the sodium concentration are the main determinant of ADH secretion. ADH increases the permeability of the renal collecting tubule to water and allows water to flow down its concentration gradient to be reabsorbed into the hypertonic medullary interstitium. Release of ADH leads to renal water retention and a decrease in the tonicity of the ECFV.

ADH release is quite sensitive. Changes of only a few mOsm/L will stimulate hypothalamic osmoreceptors and lead to ADH release. The urine osmolality may be as high as 1200 mOsm/L when ADH is present, and as low as 50 mOsm/L when ADH is absent. A number of nonosmotic stimuli can cause ADH secretion, even though ECFV tonicity is not increased. Serious ECFV depletion may "override" the osmotic control of ADH secretion. Volume depletion may cause ADH release even if the sodium concentration is normal or if hyponatremia is present. Nausea and narcotics can stimulate ADH release. There are a number of clinical disorders and drugs that can increase ADH secretion or enhance its action at the collecting tubule. These nonosmotic releasers or enhancers of ADH can cause water retention and are important causes of hyponatremia. The clinical syndrome of nonosmotic release or enhancement of ADH action leading to pathologic water retention and hyponatremia is called the **syndrome of inappropriate ADH (SIADH).**

On the other hand, absence or deficiency of ADH can lead to inability of the kidney to appropriately concentrate the urine. This concentrating defect can lead to excessive renal water losses and hypernatremia. The syndrome of ADH deficiency leading to excessive renal water loss is called **central diabetes insipidus.**

Collecting tubule responsiveness to ADH

Certain disorders are associated with tubular unresponsiveness to ADH. This tubular unresponsiveness creates a renal concentrating defect that leads to excessive renal losses of water. This syndrome is termed **nephrogenic diabetes insipidus.** Even though there are adequate levels of circulating ADH, the collecting tubule does not appropriately increase its permeability to allow water reabsorption, and this leads to excessive renal water loss and the potential for hypernatremia. Conversely, there are conditions and certain drugs that have an ADH-like effect on the collecting tubule or increase the tubular sensitivity to ADH. These conditions can lead to inappropriate water retention and hyponatremia by causing SIADH.

Thiazides, Loop Diuretics, and Hyponatremia

Both thiazide and loop diuretics block sodium reabsorption, resulting in sodium loss from the body (**Fig. 1-1**). Because sodium is the major extracellular cation and because the amount of sodium in the ECFV essentially determines the size of the ECFV, the loss of sodium is accompanied by a decrease in the size of the ECFV. Both loop diuretics and thiazide diuretics are capable of causing sodium loss and a decrease in the size of the ECFV.

Thiazide and loop diuretics differ in an important way. Loop diuretics cause greater loss of both sodium and water than do thiazides. The sodium loss from loop diuretics is greater because they block sodium reabsorption in the ascending limb of the loop of Henle, where 20–30% of filtered sodium is normally reabsorbed, whereas thiazides block sodium reabsorption in the distal tubule, where only 5–10% of filtered sodium is reabsorbed.

Because loop diuretics cause a greater loss of sodium than do thiazides, one might mistakenly think that loop diuretics should produce more hyponatremia, but the opposite is true. Remember that hyponatremia is the result of an excess of water relative to sodium in the ECF, and that hyponatremia is the result of retention of water, not the loss of sodium. Loop diuretics generally cause proportional losses of both sodium and water such that the sodium and water composition of the ECF is generally left undisturbed. The ECF sodium concentration is therefore left unchanged. Thiazides, on the other hand, may cause proportional losses of sodium and water such that a relatively *less* amount of water is excreted than sodium. This *relative* retention of water with regard to sodium can change the sodium and water composition of the ECF, lowering the ECF sodium concentration. In fact, thiazides cause hyponatremia to a degree that they are contraindicated in patients with hyponatremia.*

---

*The following optional reading is a more detailed account of why thiazides cause hyponatremia while loop diuretics generally do not.

- Normally, when one ingests large amounts of water, the kidney responds by producing large amounts of dilute urine, thereby avoiding a hyponatremic ECFV. The cortical ascending limb of the loop of Henle and the distal tubule reabsorb sodium and chloride from the tubular fluid but are impermeable to water. The reabsorption of sodium and chloride without water produces a dilute tubular fluid. The tubular fluid osmolality may be as low as 50 mOsm/L as it enters the collecting tubule. In effect, the cortical ascending limb of the loop of Henle and the distal tubule form a "diluting" segment of the nephron even though this is not a true dilution (by adding water) but a relative dilution (by subtracting sodium). When large amounts of water are ingested, ADH secretion is "turned off" and the collecting tubule is rendered impermeable to water, so water is not

12

Guidelines to Solving Clinical Problems of Sodium and Water

It is important for the body to maintain its extracellular sodium concentration within a narrow range (135–145 mEq/L) and to maintain the size of the ECFV within an acceptable range. In principle, mechanisms that control the intake and excretion of water influence the size of the ECFV to some extent, but the mechanisms that control the intake and output of sodium are far more important in determining the size of the ECFV because sodium is the major extracellular cation and contributes the osmotic driving force that maintains the ECFV. In principle, the *concentration* of ECF sodium could be affected by the amount of sodium intake or excretion, but the mechanisms that control the intake and output of water are far more important in determining the ECF sodium concentration.

- *In clinical practice, it is most useful to consider that cases of abnormal ECFV size are due to problems with the sodium control mechanisms.*
- *In clinical practice, it is most useful to consider that cases of abnormal ECF sodium concentration are due to problems with the water control mechanisms.*

---

reabsorbed into the hypertonic medullary interstitium. The dilute tubular fluid is then excreted at a concentration as low as 50 mOsm/L. This is how the kidney is able to rid the body of excess water and thereby defend the ECFV against hyponatremia. If the kidney cannot produce a dilute urine in response to a water load, the ECFV will become dilute and hyponatremia will develop.

- Thiazide diuretics interfere with the ability to produce a maximally dilute urine by blocking sodium reabsorption in the distal tubule. A patient taking thiazides may not be able to produce enough dilute urine to prevent a fall in the ECFV sodium concentration caused by ingested water. Therefore, hyponatremia may develop. A patient on thiazides is at risk for developing hyponatremia if too much water is ingested.
- Sodium reabsorption in the loop of Henle contributes to the hypertonic medullary concentration gradient that is important in water reabsorption (osmotically moving water from the collecting duct). This enables the kidney to concentrate the urine. Loop diuretics block sodium reabsorption in the loop of Henle and this blockage interferes with the maintenance of medullary hypertonicity. Therefore, loop diuretics impair water reabsorption and the ability of the kidney to *concentrate* the urine.
- Loop diuretics also interfere to some extent with urinary *dilution* by blocking the reabsorption of sodium without water in the ascending limb of the loop of Henle (**Fig. 1-1**). Therefore, loop diuretics interfere with *both* urine concentration and dilution. *The patient taking loop diuretics generally avoids the problem of hyponatremia by reabsorbing less water because of the less hypertonic medullary interstitium.*
- Because loop diuretics cause greater loss of both sodium and water and lead to a greater decrease in ECFV than do thiazides, they are preferred in edematous states. They are also the preferred agents in patients with edematous states and hyponatremia, because thiazides could worsen the hyponatremia and are therefore contraindicated.
- *Thiazides are generally contraindicated in patients with hyponatremia regardless of the underlying cause.*

13

When I see patients with increased ECFV or decreased ECFV, I first ask: How might the *sodium* control mechanisms be impaired? ECFV overload can be viewed as a state of having too much total body sodium. ECFV depletion can be viewed as a state of having too little total body sodium. The diagnosis and treatment must focus on finding and correcting the faulty sodium control mechanism.

When I see patients with hyponatremia or hypernatremia, I begin by asking: How might the *water* control mechanisms be impaired? A faulty water control mechanism results in too much water relative to sodium in cases of hyponatremia and in too little water relative to sodium in cases of hypernatremia. The diagnosis and treatment must focus on finding and correcting the faulty water control mechanism.

There are states which have *both* abnormal sodium concentration (a water control problem) and ECFV size (a sodium control problem). These states may be viewed as having abnormalities of *both* the water and sodium control mechanisms, and diagnosis and treatment must focus on finding and correcting these (see **Fig. 1-3**). You should *not* try to memorize Fig. 1-3. It is included for reference and to illustrate the many possible combinations and associated clinical conditions of abnormal ECFV (which means abnormal sodium control mechanisms) and abnormal ECF sodium concentration (which means abnormal water control mechanisms). If these concepts are not completely clear at this point, do not worry. The ideas will be expanded and developed with numerous examples and exercises.

Potassium Physiology and Pathophysiology

Potassium is the major intracellular cation. Maintenance of a stable plasma potassium concentration is essential for normal cellular function, cardiac rhythm, and proper neuromuscular transmission. The concentration of potassium in cells is roughly 130–140 mEq/L, in *marked* contrast to the extracellular concentration of only 3.5–5.0 mEq/L. Total body potassium is distributed 98% intracellularly and 2% extracellularly. Consequently, even a tiny change in this distribution could mean significant hypokalemia or hyperkalemia, even when total body potassium stores are normal.

Transcellular Potassium Distribution and Potassium Uptake by Cells

The large transcellular potassium gradient is maintained by the sodium-potassium ATPase pump located in the cell membrane. This energy-requiring pump actively transports sodium out of cells and potassium into cells in an exchange ratio of 3 sodium/2 potassium. Several important factors, both physiologic and pathologic, affect the transcellular distribution of potassium and therefore, the plasma potassium concentration:

• Insulin causes potassium to move into cells. Patients with a deficiency of insulin have impaired assimilation of potassium into cells and are at risk for developing hyperkalemia.

FIGURE 1-3. States of Abnormal ECF Volume and Abnormal ECF Sodium Concentration

Disorder(s)	Implication	Primary problem (where to start looking)	Examples of common clinical causes
Hyponatremia ECFV normal	Water excess relative to sodium	Abnormal water control (too much water relative to sodium)	SIADH
Hypernatremia ECFV normal	Water deficit relative to sodium	Abnormal water control (too little water relative to sodium)	Diabetes insipidus Insensible losses
Sodium concentration normal ECFV increased	Increased total body sodium	Abnormal sodium control (too much sodium)	CHF Cirrhosis Nephrotic syndrome Renal failure
Sodium concentration normal ECFV decreased	Decreased total body sodium	Abnormal sodium control (too little sodium)	Vomiting Diarrhea Loop diuretics
Hyponatremia with increased ECFV	Water excess relative to sodium *and* increased total body sodium	Abnormal water control (too much water relative to sodium) and abnormal sodium control (too much sodium)	CHF Cirrhosis Nephrotic syndrome Renal failure
Hyponatremia with decreased ECFV	Water excess relative to sodium *and* decreased total sodium.	Abnormal water control (too much water relative to sodium) and abnormal sodium control (too little sodium)	Vomiting Thiazide diuretics
Hypernatremia with increased ECFV	Water deficit relative to sodium *and* increased total body sodium	Abnormal water control (too little water relative to sodium) and abnormal sodium control (too much sodium)	Administration of hypertonic sodium solutions or $NaHCO_3$ (iatrogenic)
Hypernatremia with decreased ECFV	Water deficit relative to sodium *and* decreased total body sodium	Abnormal water control (too little water relative to sodium) and abnormal sodium control (too little sodium)	Osmotic diuresis Diarrhea

- pH. Changes in extracellular fluid pH can cause transcellular shifts of potassium. Acidosis tends to cause potassium ions to leave cells in exchange for hydrogen ions and therefore raises the plasma potassium concentration, whereas alkalosis does the opposite. In metabolic acidosis, much of the excess hydrogen ion is buffered intracellularly. Electroneutrality is preserved by potassium leaving the cell, which results in a variable increase in plasma potassium, depending upon the type of acidosis present. *Inorganic* acids tend to cause potassium to shift out of cells, resulting in an increase in the plasma potassium concentration. *Organic* acids such as ketoacids and lactic acid tend not to produce significant transcellular shifts in potassium for reasons that are complex.
- Stimulation of beta$_2$ adrenergic receptors causes potassium to shift into cells. This movement is mediated in part by increased activation of sodium-potassium ATPase.
- Large pathologic increases in osmolality such as occur in severe hyperglycemia can cause shifts of potassium into the extracellular fluid compartment and raise the plasma potassium concentration. The mechanism for this egress of potassium from cells is thought to be twofold: Water flows out of cells in response to the increase in ECFV tonicity, thereby raising the intracellular potassium concentration. The increased intracellular potassium concentration favors potassium movement out of cells. A second mechanism is solvent drag, whereby water carries potassium along with it through the cell membrane.

Sources of Potassium

Normally, dietary potassium intake is matched by urinary and stool potassium losses. The average diet contains about 1 mEq/kg body weight/day of potassium, which amounts to approximately 70 mEq/day in a 70 kg person. This potassium is normally excreted 90% in the urine and 10% in the stool. There are other important "hidden" sources of potassium intake that are important to remember:

- Breakdown of tissue, such as in rhabdomyolysis, hemolysis, and following chemotherapy of certain leukemias and lymphomas
- Blood transfusion
- Gastrointestinal hemorrhage with potassium absorption
- Potassium in intravenous and hyperalimentation fluids and in tube feedings
- Potassium in medications

Because of the large potassium excretory capacity of the normal kidney, hyperkalemia generally does not develop unless a renal excretory defect is also present.

Renal Potassium Excretion

By far the most important route of elimination of excess potassium is renal excretion. There is a large range of potassium excretion by the kidney. In potassium deficiency, the normal kidney can reduce daily urine potassium losses to 10 mEq/24 hours or less to conserve potassium. During increased potassium intake or endogenous potassium release from muscle, the daily potassium excretion may be as high as 10 mEq *per kg body weight*/24 hours. (For example, as high as 700 mEq/24 hours in a 70 kg person). The ability of the kidney to excrete excess potassium declines as GFR is reduced by renal failure. When the GFR is diminished below 20% of normal, the kidney has difficulty excreting the daily dietary potassium load, and hyperkalemia may develop.

Potassium is filtered freely at the glomerulus, and approximately 10% of filtered potassium reaches the collecting tubule. It is what happens in the collecting tubule that ultimately determines the amount of potassium excreted by the kidney. Potassium excretion involves the active pumping of potassium from the peritubular interstitium into the tubular cell interior by membrane-bound sodium-potassium ATPase (see **Fig. 1-4**). Sodium channels in the cell luminal membrane allow sodium to *enter* the cell from the tubule lumen down its concentration gradient. Potassium channels allow potassium to *leave* the tubular cell and to enter the lumen down its concentration gradient.

The number of functional sodium and potassium channels is determined by aldosterone. Aldosterone increases sodium-potassium exchange by binding to intracellular receptors and increasing the number of functional sodium and potassium channels. Sodium entering the tubular cell from the lumen is

COLLECTING TUBULE POTASSIUM SECRETION

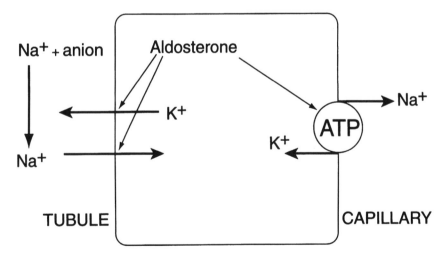

FIGURE 1-4. The amount of K^+ excreted is increased by **aldosterone** which opens Na^+ and K^+ channels and stimulates Na^+-K^+ ATPase, and by **Na^+ delivery** which can be increased by diuretics, saline infusion, and filtration of poorly reabsorbed anions such as excess HCO_3^-, which "carry" Na^+ to the collecting tubule. Large K^+ losses result when *both* **aldosterone** and increased **Na^+ delivery** are present.

pumped out of the cell by the basolateral sodium-potassium pump and returns to the ECF. The potassium that has entered the tubule lumen is excreted. The net result is sodium reabsorption and potassium secretion, sometimes referred to as sodium-potassium exchange. Four important factors influence distal sodium-potassium exchange and serve to control the final amount of potassium in the urine. These are clinically important mechanisms to remember when considering disorders of potassium:

- Aldosterone causes increased exchange of sodium-potassium and increases the amount of potassium in the urine, depending upon the amount of sodium/volume delivered to the distal nephron. Aldosterone is stimulated by activation of the renin-angiotensin system and by hyperkalemia. Aldosterone excess results in renal potassium loss and hypokalemia, whereas aldosterone deficiency results in renal potassium retention and hyperkalemia.
- Increased delivery of sodium to the collecting tubule increases the potassium excretion rate by increasing the amount of sodium presented for exchange with potassium. For example, loop and thiazide diuretics, osmotic diuresis, and saline infusions increase delivery of sodium to the collecting tubule and cause increased potassium excretion.
- The presence of a poorly reabsorbable anion. For example, during metabolic alkalosis, excess HCO_3^- that cannot be reabsorbed by the proximal tubule "carries" sodium to the collecting tubule as an accompanying cation. This increased delivery of sodium increases sodium-potassium exchange and increases urinary potassium excretion.
- Acid-base status: Acidosis inhibits potassium secretion and alkalosis increases potassium secretion.

Large renal potassium losses may occur when *both* increased sodium delivery to the collecting tubule *and* high aldosterone levels are present. For example, in diabetic ketoacidosis, osmotic diuresis leads to increased delivery of sodium to the collecting tubule and also causes ECFV depletion, which stimulates aldosterone. Together, high aldosterone levels and increased sodium delivery to the collecting tubule lead to large urinary potassium losses. Severe potassium depletion can occur in diabetic ketoacidosis.

On the other hand, aldosterone deficiency and tubular unresponsiveness to aldosterone are important causes of hyperkalemia due to decreased potassium excretion.

Extrarenal Potassium Loss

Losses of potassium in sweat are usually minimal: Sweat contains around 9 mEq/L potassium and sweat volume is about 200 ml/24 hours in a sedentary person. The daily sweat loss of potassium is therefore only about 9 mEq/L × .2 L = 1.8 mEq per day. During vigorous activity and in hot climates, however, sweat volume may reach 10 L/day and potassium losses may be as high as 9 mEq/L $\times$ 10 L = 90 mEq/ day!

Stool losses of potassium are normally about 10% of the dietary potassium load, but much larger losses of potassium may occur with diarrhea. Stool losses of potassium increase in chronic renal failure as the body defends itself against hyperkalemia.

Hydrogen Ion Physiology and Pathophysiology

The precise control of blood pH within the narrow range of 7.35–7.45 is accomplished by regulation of hydrogen ion balance. The hydrogen ion concentration is dictated by the ratio of two quantities: the HCO_3^- concentration, which is regulated by the kidneys, and the P_{CO2}, which is controlled by the lungs. This relationship is expressed by:

$$CO_2 + H_2O \leftrightarrow H_2CO_3 \leftrightarrow HCO_3^- + H^+$$
$$[H^+] \propto (P_{CO2}/HCO_3^-])$$

This equation describes the mass action of the CO_2-HCO_3^- buffering system, which is the main buffering system in extracellular fluid. The hydrogen ion concentration is determined by the *ratio* of the P_{CO2} and the HCO_3^- concentration ($[HCO_3^-]$). Notice that the hydrogen ion concentration may increase from either an increase in P_{CO2} (respiratory acidosis) or a decrease in $[HCO_3^-]$ (metabolic acidosis). The hydrogen ion concentration may decrease by either a decrease in P_{CO2} (respiratory alkalosis) or an increase in $[HCO_3^-]$ (metabolic alkalosis). Normally, the lungs keep the P_{CO2} in the range of 40 mm Hg, and the kidneys keep the HCO_3^- concentration between 24–26 mEq/L. The lungs and the kidneys defend the body pH against hydrogen ion gain or loss.

When one HCO_3^- is lost from the body, one hydrogen ion stays behind. The net result is the addition of one free hydrogen ion to the body. Therefore, the loss of a HCO_3^- has the *same* result as the gain of one hydrogen ion. Conversely, the gain of one HCO_3^- is the *same* as the loss of one hydrogen ion.

Under normal conditions, there are two sources of hydrogen ion that the body must eliminate:

- About 20,000 mmols of CO_2 are produced each day by the metabolism of carbohydrates and fats. This large amount of CO_2 is eliminated by the lungs. Although CO_2 is not an acid, it combines with H_2O to form H_2CO_3; therefore, acid would accumulate very rapidly if CO_2 were not adequately excreted by the lungs (respiratory acidosis).
- About 1 mEq/kg (50–100 mEq) of nonvolatile acid is produced *each day* by the metabolism of protein. This hydrogen ion is called the "fixed" acid

load because it cannot be eliminated by the lungs. It is buffered by the HCO_3^- in extracellular fluid. This utilization of HCO_3^- to buffer the daily 50–100 mEq of hydrogen ion would lead to HCO_3^- depletion and metabolic acidosis except for the kidney's ability to generate new bicarbonate. The kidney makes new HCO_3^- by eliminating hydrogen ion from the body, which adds one HCO_3^- to extracellular fluid for every hydrogen ion eliminated. Remember, when one hydrogen ion leaves the body, it leaves behind one HCO_3^-. Renal elimination of 50–100 mEq/day of hydrogen ion keeps the $[HCO_3^-]$ within its narrow range of 24–26 mEq/L.

Body Buffers

Body buffer systems are the first line of defense against acute changes in hydrogen ion concentration. Hydrogen ions are buffered by both intracellular and extracellular buffers. Intracellular buffers include phosphates and cytosolic proteins. The main extracellular buffer is the CO_2-HCO_3^- system. This is the body buffering system that we assess in the clinical lab when we order an "arterial blood gas" test, which measures P_{O2} (arterial partial pressure of oxygen in mm Hg), P_{CO2} (arterial partial pressure of carbon dioxide in mm Hg), and pH. An arterial blood gas test generally also gives a value for the HCO_3^- concentration (in mEq/L), which is calculated using the Henderson-Hasselbalch equation using the measured pH and P_{CO2}.

Renal Regulation of the HCO₃⁻ Concentration

The kidney regulates $[HCO_3^-]$ by two very different means. Both are necessary to maintain $[HCO_3^-]$ within normal limits (24–26 mEq/L).

Reabsorption of virtually all of the filtered HCO_3^- by the proximal tubule

Almost all of the HCO_3^- filtered by the glomerulus is reclaimed by the proximal tubule (see **Fig. 1-5**). This process is a high capacity system because a huge amount of HCO_3^- is filtered by the kidney each day and requires reclamation:

$$180 \text{ L/day (GFR)} \times 25 \text{ mEq/L (filtered concentration)}$$
$$= 4500 \text{ mEq/day of } HCO_3^- \text{ that must be reclaimed!}$$

This process does not add net HCO_3^- to the ECFV nor secrete net hydrogen ion into the urine. It does *nothing* to change the acid-base status of the body: total hydrogen ion does not change. This process simply keeps HCO_3^- from being lost in the urine and therefore prevents metabolic acido-

PROXIMAL TUBULE HCO₃⁻ RECLAMATION

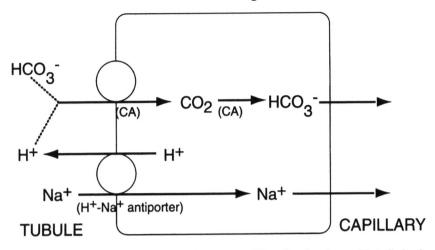

FIGURE 1-5. The Na^+-H^+ antiporter secrets one H^+ and reabsorbs one Na^+. Carbonic anhydrase (CA) in the proximal tubular lumen brush border catalyzes $H^+ + HCO_3^-$ to CO_2 and H_2O). CO_2 diffuses into the cell where intracellular CA catalyzes $CO_2 + OH^-$ to HCO_3^-. The net result is reabsorption of 1 $NaHCO_3$. A defect in this system leads to proximal (type II) renal tubular acidosis.

sis from developing. HCO_3^- reabsorption is normally complete at a filtered (plasma) concentration of 24–26 mEq/L or less. Above this "threshold" concentration of 24–26 mEq/L, however, the amount of HCO_3^- reaching the proximal tubule becomes greater than the ability of the proximal tubule to reclaim HCO_3^-, and the system is overwhelmed. The proximal tubule can no longer reclaim all the filtered HCO_3^-, and the HCO_3^- that is not reabsorbed begins to "spill" into the urine. Consequently, the elevated plasma HCO_3^- tends to return toward the threshold value of $[HCO_3^-]$. Several important factors increase the rate of proximal reabsorption of HCO_3^-:

- ECFV status. A decrease in ECFV increases the proximal reabsorption of HCO_3^-. Because ECFV depletion leads to increased HCO_3^- reabsorption, ECFV depletion is an important factor in sustaining an elevated HCO_3^- concentration in patients with metabolic alkalosis. The increased proximal reabsorption of $NaHCO_3$ sustains the metabolic alkalosis until the ECFV depletion is corrected.
- Increased angiotensin II increases proximal reabsorption of $NaHCO_3$.
- An increase in P_{CO2} results in an increase in proximal reabsorption of HCO_3^- and a higher plasma HCO_3^-. A decrease in P_{CO2} does the opposite. This is important in the kidney's compensatory response to respiratory acidosis (increased P_{CO2}) and respiratory alkalosis (decreased P_{CO2}).

- Severe depletion of potassium stores from any cause increases proximal re-absorption of HCO_3^-. The mechanisms are complex. Hyperkalemia does the opposite.

When a defect develops in the proximal tubular reabsorption of filtered HCO_3^-, the HCO_3^- concentration falls as HCO_3^- is lost in the urine. This lowering of the HCO_3^- concentration results in metabolic acidosis. The syndrome of metabolic acidosis caused by defective proximal tubular HCO_3^- reabsorption is called **proximal (type II) renal tubular acidosis.**

Renal excretion of hydrogen ion

The second way the kidney controls the plasma $[HCO_3^-]$ is by eliminating enough hydrogen ion to equal the fixed acid produced each day (see **Fig. 1-6**). Remember, the removal of one hydrogen ion is equivalent to the *gain* of one HCO_3^-. The removal of hydrogen ion from the body by the kidney results in the generation of "new" HCO_3^- to replace the 50-100 mEq/day of HCO_3^- that was used to buffer the daily production of fixed acid. The kidney does this by two mechanisms:

- Active secretion of hydrogen ion by an ATP-utilizing proton "pump" in the collecting tubule. One HCO_3^- is produced for every hydrogen ion excreted.
- Hydrolysis of glutamine in the proximal tubule generates NH_4^+ (which is excreted in the urine) and HCO_3^- (which is returned to the ECFV). The process

COLLECTING TUBULE HYDROGEN ION EXCRETION
(= BICARBONATE REGENERATION)

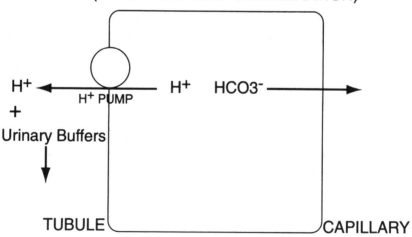

FIGURE 1-6. The hydrogen ion "pump" in the collecting tubule secretes one H^+ which is excreted. This has the net effect of adding one HCO_3^- to ECF. A defect in this system may lead to distal (type I) renal tubular acidosis.

of ammoniagenesis rids the body of hydrogen ion so long as the NH_4^+ ions produced are excreted in the urine. The precise mechanisms of NH_4^+ generation and excretion are complex and are not detailed here.

NH_4^+ excretion is quantitatively more important than secretion of hydrogen ion in generating HCO_3^-. The main way the kidney responds to acidosis (excess hydrogen ion) is by increasing the production and excretion of NH_4^+.

When a defect develops in the renal elimination of H^+, the HCO_3^- concentration falls as HCO_3^- is used to titrate the excess H^+ produced each day. This lowering of the HCO_3^- concentration results in a metabolic acidosis. The syndrome of metabolic acidosis caused by defective renal tubular elimination of hydrogen ion is called **distal (type I) renal tubular acidosis.**

The Anion Gap

Calculation of the anion gap is essential to analyzing acid-base disorders correctly. The extracellular fluid is electroneutral: the sum of the concentrations of the positively charged ions must equal that of the negatively charged ones. This concept can be expressed by the equation:

$$Na^+ + UC = Cl^- + HCO_3^- + UA$$

where UC (unmeasured cations) indicates the sum of the charges of all cations other than sodium, and UA (unmeasured anions) equals the sum of the charges of all anions other than chloride and bicarbonate. The major UC are potassium, calcium, magnesium, and some gamma globulins. The major UA are albumin, sulfate, phosphate, and various organic anions. The above equation can be rearranged to derive the expression for the **anion gap (AG):**

$$AG = UA - UC = [Na^+] - ([Cl^-] + [HCO_3^-])$$

The AG is normally 9–16 mEq/L. This normal range depends upon the individual hospital, however. Many hospitals may prefer to use a smaller normal range of 10–14, which is approximately ± 1 standard deviation. The wider range 9–16 mEq/L is closer to ± 2 standard deviations. Some types of metabolic acidosis add hydrogen ion along with an associated unmeasured anion to the ECF. The addition of the acid H-Anion affects both sides of the equation for the AG. The hydrogen ion is buffered by HCO_3^- and therefore lowers the HCO_3^- concentration, and the anion increases the AG by adding to the unmeasured anions (UA). The result is a so-called high anion gap acidosis.

In clinical practice, we separate metabolic acidosis into two categories: high anion gap and normal anion gap (also called non anion gap or hyperchloremic). If the anion gap is increased to the range of 30 mEq/L or more, then a high anion gap metabolic acidosis is virtually always present *regardless* of what the pH and the $[HCO_3^-]$ are. If the anion gap is increased to the range 20–30 mEq/L, then it is likely that a high anion gap metabolic acidosis is present *regardless* of what the pH and the $[HCO_3^-]$ are. Acid-base disorders are discussed in Chapters 7, 8 and 9.

Exercises

The exercises at the end of each chapter are intended to expand on the text and to introduce new material in the context of clinical cases.

1. Estimate the total body water in a 50 kg woman.
   Answer: .5 × 50 = 25 liters. In an elderly woman, the total body water would be less (perhaps 20 liters).

2. Estimate the total body water in a 100 kg man.
   Answer: .6 × 100 = 60 liters. This is more than twice the total body water of the 50 kg woman. In an elderly man, the total body water would be less (perhaps 50 liters).

3. Estimate the ECFV in a 50 kg woman.
   Answer: Total body water: .5 × 50 = 25 liters. The ECFV is approximately 1/3 of total body water: 25/3 = 8.3 liters.

4. Estimate the ECFV in a 100 kg man.
   Answer: Total body water: .6 × 100 = 60 liters. The ECFV is approximately 1/3 of total body water: 60/3 = 20 liters.

5. Estimate the total ECF sodium in a 50 kg woman.
   Answer: Total body water: .5 × 50 = 25 liters. The ECFV is approximately 1/3 of total body water: 25/3 = 8.3 liters. Now multiply by the sodium concentration in extracellular fluid (normally around 140 mEq/L): 8.3 L × 140 mEq/L = 1162 mEq.

6. Estimate the total ECF sodium in a 100 kg man.
   Answer: Total body water: .6 × 100 = 60 liters. The ECFV is approximately 1/3 of total body water: 60/3 = 20 liters. Now multiply by the sodium concentration in extracellular fluid (normally around 140 mEq/L): 20 L × 140 mEq/L = 2800 mEq.

7. Review the answers to exercises 1–6. Notice the large differences between the 50 kg woman and the 100 kg man. The first rule of clinical electrolyte and acid-base physiology is that *most patients are not "standard" 70 kg men.* This is especially important to remember when calculating electrolyte replacement and in planning IV fluid therapy.

8. A patient has the following chemistries: sodium 140 mEq/L, glucose 180 mg/dl, BUN 28 mg/dl. What are the contributions of each of these three constituents to serum osmolality?
   Answer:
   Sodium (along with chloride and other anions) contributes 2 × 140 = 280 mOsm/L.
   Glucose contributes 180/18 = 10 mOsm/L.
   Urea contributes 28/2.8 = 10 mOsm/L.

9. A patient with renal failure has the following chemistries: sodium 130 mEq/L, glucose 100 mg/dl, BUN 120 mg/dl. Calculate the osmolality. Would you expect this osmolality to be associated with increased thirst?
Answer: The calculated osmolality is $2 \times 130 + 100/18 + 120/2.8 = 308$. The increase in osmolality is due to an increase in urea, which is not an effective osmole. Therefore, this patient's ECF is not *hypertonic* and thirst would not be stimulated.

10. A patient with diabetes has the following chemistries: sodium 140 mEq/L, glucose 900 mg/dl, BUN 28 mg/dl. What is the osmolality? Is this patient hypertonic?
Answer. The calculated osmolality is 340. Yes, this patient is hypertonic, because glucose is an effective osmole and is capable of translocating water.

11. What is the contribution of glucose to the osmolality in the previous example?
Answer: $900/18 = 50$ mOsm/L.

12. A patient presents to the emergency room with the following chemistries: sodium 140 mEq/L, glucose 360 mg/dl, BUN 28 mg/dl, measured osmolality 360. Calculate the osmolal gap. What are the substances that can cause an increase in the osmolal gap? How much is the glucose contributing to the increased osmolal gap?
Answer: The calculated osmolality is $2 \times 140 + 360/18 + 28/2.8 = 310$. The osmolal gap is 50 (normal is less than 10). Exogenous substances that can cause an increase in the osmolal gap are: mannitol, ethanol, isopropanol, methanol, ethylene glycol, and sorbitol. We would need more information to decide which of these is the culprit. The glucose is contributing $360/18 = 20$ to *both* the calculated and the measured osmolalities and therefore does not affect the osmolal gap.

13. A 42-year-old patient presents with a sodium concentration of 120 mEq/L. What can you say about the mechanisms of water regulation in this patient? What is the status of total body sodium?
Answer: The low serum sodium concentration tells us that water regulation is abnormal. For clinical purposes, disorders of the sodium *concentration,* both hyponatremia and hypernatremia, can be viewed as originating from abnormalities in the regulation of water homeostasis. The abnormal sodium concentration leads us to start our investigation with the question: Why is water regulation abnormal? In the case of hyponatremia, there is too much water relative to sodium because the kidney is not properly excreting water. The sodium *concentration* does not tell us anything reliable about whether total body sodium is increased, decreased, or normal.

We are given no clinical information about the size of the ECFV. Therefore, we cannot say anything about whether total body sodium is increased, decreased, or normal. To assess total body sodium, we must clinically assess the size of the ECFV, because the size of the ECFV is determined by the amount of total body sodium. Signs of ECFV depletion, indicating total body sodium depletion, are poor skin turgor, dry mucosa, orthostatic fall in blood pressure, and orthostatic rise in heart rate. Signs of ECFV overload, indicating total body sodium overload, are jugular venous distention, pulmonary rales, pleural effusion, ascites, S3 gallop, and, of course, pretibial edema.

14. A patient presents with massive pedal edema and ascites. His sodium concentration is 140 mEq/L. Does he have a problem with sodium control, water control, or both?
    Answer: This patient has a problem with sodium control. For the purposes of solving clinical problems, it is useful to consider that abnormalities of the size of the ECFV result from an abnormal amount of total body sodium. Clinical assessment of the ECFV tells us roughly whether total body sodium is increased, decreased or normal. This patient has a markedly expanded ECFV, as indicated by the pedal edema and ascites. Therefore, total body sodium is markedly increased. The edema-forming states, congestive heart failure, cirrhosis with ascites and edema, and nephrotic syndrome, can be viewed as states having too much total body sodium. Therefore, sodium control is abnormal.

    The sodium *concentration* tells us about water control. If the sodium *concentration* is normal, then there is no *clinically significant* problem with water control. This patient may have trouble excreting a water load and may develop hyponatremia if given large amounts of water, but the sodium concentration is normal: therefore, he does not presently have a *clinically significant* disorder of water regulation. *An important clinical point is that it is unnecessary to restrict water in a patient with congestive heart failure, cirrhosis, or nephrotic syndrome if the serum sodium concentration is normal.*

15. A 34-year-old man has a sodium concentration of 125 mEq/L. What can you say about the status of his total body sodium? Is he sodium depleted?
    Answer: We cannot say anything reliable about the status of this patient's total body sodium based on the low serum sodium concentration. We are given no clinical information about the status of the ECFV so we cannot say anything about the status of total body sodium. The sodium concentration is just that—a concentration, not a measure of total amount. The sodium concentration by itself does not tell us anything reliable about the status of total body sodium. The sodium concentration does not tell us whether total body sodium is increased, decreased, or normal. Total body sodium can be approximated by clinical assessment of the size of the ECFV because the *amount* of total body sodium is the main determinant of the size of the ECFV.

    This patient has a problem with *water* control because the serum sodium concentration is abnormal. The sodium concentration of 125 mEq/L tells us that there is an excess of water relative to sodium in the ECFV and therefore that there is abnormal regulation of *water.*

Briefly: abnormal ECFV means there is an abnormal amount of total body sodium. Abnormal sodium concentration means there is abnormal water regulation.

16. A 23-year-old man presents with pedal edema, rales, and a third heart sound (S3). His sodium concentration is 120 mEq/L. Does he have a problem with sodium regulation, water regulation, or both?

Answer: Both. The ECFV is increased (disorder of sodium regulation) *and* the serum sodium is abnormal (disorder of water regulation). This patient has a serious problem with sodium regulation. He has an expanded ECFV (edema, rales, S3) caused by congestive heart failure and therefore has an excess of total body sodium. The hyponatremia tells us that this patient also has a serious problem with water regulation. The low serum sodium means that there is excess of water relative to sodium. This patient has an impaired ability to excrete enough dilute urine to get rid of excess water. This patient needs loop diuretics and sodium restriction to reduce the total body sodium, *and* he needs water restriction because of the hyponatremia.

17. A patient presents with poor skin turgor and tachycardia. His blood pressure falls with standing up. His sodium concentration is 130 mEq/L. Does he have a problem with sodium regulation, water regulation, or both?

Answer: Both. Clinical assessment of the ECFV tells us roughly whether total body sodium is increased, decreased or normal. This patient has a decreased ECFV. Therefore, total body sodium is decreased. The clinical ECFV depletion is due to a large decrease of total body sodium. The hyponatremia is due to abnormal renal water retention; there is water excess relative to sodium because the kidney cannot produce enough dilute urine to get rid of excess water in the presence of ongoing water intake. This patient has clinically important disorders of *both* sodium and water regulation. Both the hyponatremia and the depleted ECFV should respond to replacement of the ECFV with isotonic NaCl solution. IV fluids will be covered in Chapter 2.

18. Which of the following will increase the measured serum osmolality when added to the extracellular fluid?

Urea ✓
Glucose ✓
Sodium ✓
Ethanol ✓
Methanol ✓
Isopropanol ✓
Ethylene glycol ✓
Mannitol ✓
Sorbitol ✓

Answer: They will all increase the measured osmolality if added to the extracellular fluid.

19. Which of the following will increase the <u>calculated</u> serum osmolality when added to the extracellular fluid?

Urea ✓
Glucose ✓
Sodium ✓
Ethanol
Methanol
Isopropanol
Ethylene glycol
Mannitol
Sorbitol

Answer: Only urea, glucose and sodium are included in the formula for calculated osmolality. Therefore, only urea, glucose and sodium will add to the calculated osmolality if added to extracellular fluid.

20. Which of the following will increase the osmolal gap when added to the extracellular fluid?

Urea
Glucose
Sodium
Ethanol ✓
Methanol ✓
Isopropanol ✓
Ethylene glycol ✓
Mannitol ✓
Sorbitol ✓

The ones not in the both other equations.

Answer:

$$\text{OSM GAP} = \text{OSM}_{(meas)} - \text{OSM}_{(calc)}$$

Urea, glucose, and sodium are all included in the formula for calculated osmolality. They will add to both the calculated osmolality and the measured osmolality and therefore will not change the osmolal gap if added to the extracellular fluid. The other compounds will increase the measured osmolality but not the calculated osmolality and will therefore increase the osmolal gap.

21. At approximately what level of GFR would a patient have problems excreting the daily dietary potassium load? At this point, the patient will begin to develop positive potassium balance, leading to hyperkalemia.
Answer: The upper limit of potassium excretion is *roughly* proportional to the GFR. If the GFR is 100% of normal, the maximum amount of potassium which could be excreted in one day is roughly 10 mEq per kg body weight. This is about $70 \times 10 = 700$ mEq in a 70 kg person. If the GFR is reduced to 50% of normal the maximum amount of potassium that can

28

be excreted in one day falls to approximately $50\% \times 700 = 350$ mEq. This is a rough approximation of maximum potassium excretion because compensatory renal potassium secretory mechanisms will increase potassium excretion, and stool potassium losses also increase as the body defends itself against hyperkalemia. If the GFR is further reduced to 20% of normal, the maximal potassium excretion would fall to the range of about 140 mEq/day (20% of 700 mEq/day).

The average diet has about 1 mEq of potassium per kg body weight, which amounts to about 70 mEq/day in a 70 kg person. For a diet containing 70 mEq/day, the GFR would need to be reduced to approximately $70/700 = 10\%$ of normal before hyperkalemia develops. In fact, the GFR is usually below this level when hyperkalemia develops based upon usual dietary intake. Hyperkalemia may develop at less profound levels of renal failure if the potassium intake is increased or if there is a hidden potassium load. For example, a person with a diet high in potassium would develop hyperkalemia with less impairment of the GFR. A patient with a GFR 15% of normal would develop hyperkalemia if dietary potassium is over the range of $15\% \times 700 = 105$ mEq/day. As mentioned above: this is only a rough approximation of maximum potassium excretion.

The clinical point is that if a patient has mild to moderate renal failure and hyperkalemia, the hyperkalemia should not be simply ascribed to renal failure alone. A vigorous search for other causes of hyperkalemia is needed.

22. How much potassium is there in the ECFV of a 70 kg man?
    Answer: The very delicate nature of the transcellular distribution of potassium is illustrated by the following calculation:
    TBW = $.6 \times 70$ kg = 42 L
    ECFV = $1/3 \times 42$ L = 14 L
    Potassium concentration in ECFV: 4.0 mEq/L
    Total potassium in ECFV: 4.0 mEq/L $\times$ 14 L = 56 mEq
    The calculated amount of potassium in the entire ECFV (56 mEq) is less than that contained in three routine supplemental 20 mEq doses of KCl or the potassium in four glasses of orange juice! Even a small increase in the amount of extracellular potassium could cause a large increase in the ECF potassium concentration. Adding 56 mEq to the ECFV would result in an increase of potassium concentration from 4.0 mEq/L to 8.0 mEq/L! Thankfully, we do *not* double our potassium concentration after four glasses of orange juice because homeostatic mechanisms maintain the striking difference between intracellular and extracellular potassium concentrations and, therefore, the ECFV potassium concentration.

23. How much potassium is there in the ECFV of a 40 kg woman?
    Answer:
    TBW = $.5 \times 40$ kg = 20 L
    ECFV = $1/3 \times 20$ L = 6.7 L

Potassium concentration in ECFV: 4.0 mEq/L
Total potassium in ECFV: 4.0 mEq/L $\times$ 6.7 L = 26.8 mEq
The calculated amount of potassium in the entire ECFV is about *one* supplemental 20 mEq dose of KCl!

24. Calculate the total amount of HCO_3^- present in the ECFV of a 50 kg woman with an ECF HCO_3^- concentration of 25 mEq/L.
Answer: Total body water: .5 $\times$ 50 = 25 liters. The ECFV is approximately 1/3 of total body water: 25/3 = 8.3 liters. The normal ECF stores of HCO_3^- are 25 mEq/L $\times$ 8.3 L = 207.5 mEq! This corresponds to about four standard ampules of sodium bicarbonate.
How much HCO_3^- is being reabsorbed each day by the proximal tubule assuming a GFR of 100 ml/min?
Answer: Total amount filtered = total amount reabsorbed by the proximal tubule: 100 ml/min $\times$ 1440 min/day $\times$ 25 mEq/L = 3600 mEq/day! This is about 17 times the total amount of bicarbonate in the ECF.

25. To illustrate one aspect of the importance of the urinary buffers the following is a calculation of what the urine pH would be if there were no urinary buffers. I don't expect you to know how do this calculation. It is included for illustration only.
Normally, the daily excretion of hydrogen ion is approximately 50–100 mmol/day and is equal to the amount of fixed acid produced by the metabolism of the diet. Assuming a hydrogen ion excretion of 100 mmol in a 24-hour urine volume of, say, 1 L, this would result in a urinary pH of

$$pH = -\log(H^+) = -\log(100 \text{ mmol/1L}) = -\log(.100 \text{ mmol/ml}) = 1$$

It hurts just to imagine urine with a pH of 1! Compare this pH of 1 to the normal minimum urinary pH of 4.5. The urinary buffers allow for large increases in hydrogen ion excretion (200–300 mmol/day) in states of increased hydrogen ion load without appreciable decreases in urine pH. The primary means by which the kidney rids the body of excess hydrogen ion is by increasing renal ammoniagenesis. In situations when excess hydrogen ion is added to the body, the kidney responds by increasing production and excretion of NH_4^+.

CHAPTER 2. IV SOLUTIONS AND IV ORDERS

On the clinical wards one is confronted with a constellation of different bags and bottles, each containing fluid with a strange name such as 0.9% Saline or D5 0.45% Saline. What do these solutions contain, and what are they used for? Each fluid has its own special uses and indications. This chapter tries to provide a *general* approach to the question: Which solution for which situation?

The most commonly used IV solutions are summarized in **Fig. 2-1.** A few general comments:

1) Sodium chloride (saline) solutions that have tonicities close to that of plasma are termed **isotonic.** Common examples are 0.9% saline and Ringer's Lactate. These solutions are used when it is desired to expand the extracellular fluid volume (ECFV). It is generally best to use *isotonic* rather than *hypotonic* fluids to expand the ECFV. Fluids such as D5W (5% dextrose in water), 0.45% saline and D5 0.45% saline deliver free water. Free water given in states of ECFV depletion can lead to dangerous hyponatremia.

The 5% dextrose-containing isotonic saline solutions, D5 0.9% saline and D5 Ringer's Lactate deliver a small amount of glucose (50 grams/liter). Under normal circumstances, the glucose is assimilated into cells and does not change the glucose concentration of the patient. For example, if we give 1 liter of D5 0.9% saline, we are delivering 0.9% saline to the ECFV of the patient and 50 grams of glucose, which is taken up into cells. The net result to the ECF is the addition of roughly 1 liter of 0.9% saline. However, in the patient with diabetes mellitus, the glucose is not assimilated into cells well and therefore hyperglycemia may develop with D5-containing solutions.

Some examples of situations in which 0.9% saline would be appropriate include:

- ECFV depletion from any cause—hypotonic fluids can produce dangerous hyponatremia in the setting of ECFV depletion.
- Postoperative fluid management—hypotonic fluids can produce dangerous hyponatremia in the postoperative setting.
- Shock from any cause

31

FIGURE 2-1. Electrolyte Content of Some Common IV Solutions

Solution	Glu (gm/L)	Osm	Na⁺ (mEq/L)	Cl⁻ (mEq/L)	Indications/Use	Cautions
D5W	50	252	0	0	To give free water. Small infusions (100 ml) to give a variety of medications. Does not contain sodium so will not generally produce ECFV overload.	Contains glucose: can impair control of diabetes.
0.45% NaCl	0	154	77	77	To provide both free water and sodium. Treatment of hypertonic ECFV depleted states.	Hypotonic to plasma. Can cause serious hyponatremia.
0.9% NaCl	0	308	154	154	To provide ECFV replacement. Perioperative fluid.	May cause ECFV overload in patients with CHF or renal failure.
Ringer's Lactate	0	272	130	109	To provide ECFV replacement. Perioperative fluid.	May cause ECFV overload in patients with CHF or renal failure.
3% NaCl	0	1026	513	513	Treatment of severely symptomatic hyponatremia.	Osmotic demyelinization syndrome; ECFV overload; iatrogenic hypernatremia.

- Hemorrhage
- In conjunction with blood transfusion—hypotonic fluids may cause lysis of red blood cells.
- Burns

2) Hypotonic saline solutions such as 0.45% saline can be considered to be made of approximately 1/2 0.9% normal (isotonic) saline and 1/2 water. They are generally used in situations where it is desired *both* to expand the ECFV *and* to deliver free water to a hypertonic patient. Such a patient is *both* volume-depleted *and* significantly hypertonic (usually either hypernatremic or markedly hyperglycemic or both). The sodium in the solution expands the

hyper osmolar / ECFV depleted

32

ECFV, and the water corrects the hypertonicity. Hypotonic fluids deliver free water, which can lead to hyponatremia, and therefore the serum sodium must be closely monitored. Here are some instances in which a hypotonic saline solution might be appropriate:

- Hyperosmolar states due to *severe* hyperglycemia (0.45% saline, not D5 0.45% saline)
- Hypernatremia with ECFV depletion

3) D5W is used to provide free water and is useful in the treatment of severe hypernatremia so long as it does not produce glucosuria. One liter of D5W delivers 1 liter of water to the patient, which will distribute between the ECFV and the ICFV, and 50 grams of glucose, which is normally taken up by cells. The net result is the delivery of 1 liter of free water. Pure water cannot be given intravenously because it causes hemolysis. D5W is frequently used to administer medications. One advantage of D5W is that it does not deliver unwanted sodium and therefore causes ECFV overload less readily than do saline solutions. D5W may be given at a low rate (10–25 cc/hr) when it is desired to "keep a vein open" (KVO) for intravenous medications.

Some situations in which D5W might be used are:

- Correction of hypernatremia—watch the patient carefully for hyperglycemia or glucosuria
- Delivery of medications in a non-diabetic patient
- As KVO in states of ECFV overload—D5W contains no sodium and will not further expand the ECFV as much as will saline solutions

4) Potassium supplementation is best given orally when feasible. Intravenous administration of potassium may be given

- In patients with profound, life-threatening hypokalemia
- In patients who are unable to tolerate potassium by mouth
- As a carefully chosen maintenance dose to be added to the IV fluids

Intravenous administration of potassium is potentially dangerous because of the risk of acute hyperkalemia. (Remember the delicate balance between intracellular and extracellular potassium.) Potassium is irritating to veins, and concentrations more than 30 mEq/L and rates of administration more than 10 mEq/hr are generally not recommended in nonemergency conditions.

5) One of the most useful but often overlooked measurements in medicine is the patient's weight. *Any patient who is receiving IV fluids should be weighed on a daily basis if possible.* An abrupt increase or decrease in weight is an important clue to changes in fluid status.

6) In general, daily electrolytes, blood urea nitrogen (BUN), and creatinine (Cr) should be measured in any patient receiving IV fluids to monitor therapy. In situations in which fluids are given rapidly or electrolyte

imbalances are severe, the electrolytes, BUN, and Cr should be measured more frequently.

Writing "Maintenance" IV Orders

The writing of IV orders is an important part of the everyday care of patients. One of the most important points to remember is that fluid and electrolyte therapy must be tailored to the individual patient after careful consideration of the patient's age, gender, and body mass. I always try to remember that *most patients are not "standard" 70 kg men.* There are many ways to write IV orders. This section offers some *rough guidelines* to help the beginner develop a systematic approach to writing IV orders. The following discussions assume that there are no underlying water, electrolyte, or acid-base disorders present; that there has been no recent surgery or medical illness; and that the patient has normal renal and cardiac function.

Water

Under normal circumstances, the daily requirement for water is about 2000–2500 cc per day. This requirement allows for approximately 500–1000 cc per day of loss from lungs, skin, and stool, plus about 1500 cc per day for urine volume. The patient with normal urine concentrating ability is able to excrete the daily solute load in as little as 500 cc, but there is no point in trying to minimize the urine volume. Normally, stool loss of water is less than 150 cc per day. Water requirements may be significantly more than 2000–2500 cc per day in states of fever, mechanical ventilation, or gastrointestinal losses. With fever, the ongoing insensible water loss increases by *roughly* 60–80 ml/24 hours for each degree Fahrenheit.

Sodium

The kidney can adapt to a wide range of sodium intake by either conserving or excreting sodium. In states of sodium depletion, the urine sodium may fall to less than 5 mEq/L. Therefore, it is not necessary to replace large amounts of sodium when providing "maintenance" fluids. It is customary to supply 50–100 mEq/day of sodium as sodium chloride, although patients with renal disease, congestive heart failure, or cirrhosis should receive as little sodium as possible.

Potassium

The normal kidney can also adapt to wide changes in potassium intake. In states of potassium deficiency, renal potassium excretion may be as little as 10 mEq per day. The daily diet usually contains about 1 mEq/kg per day (for example, 50 mEq per day in a 50 kg woman). Under normal circumstances

20–60 mEq/day is supplied in "maintenance" IV solutions. Administration of saline solutions without potassium supplementation can result in increased distal delivery of sodium and increased sodium-potassium exchange. This can lead to increased potassium loss in the urine, causing hypokalemia. Again, close monitoring of therapy is always indicated.

A word of caution: *The amounts of water, sodium, and potassium mentioned in this section are rough guidelines only.* Individual therapy must be tailored carefully for each patient and reassessed on a daily basis. In general, body weight, electrolytes, BUN, and creatinine should be measured daily in any patient receiving IV fluids. Also, common conditions such as renal insufficiency, congestive heart failure, and liver disease will markedly change the appropriate fluid therapy for a patient.

Exercises

Choose the best IV solution for each of the following situations:

1. A non-diabetic patient with chest pain being transferred to the coronary care unit. Vital signs are stable.
   Answer: D5W KVO for medications. An alternative for D5W is a heparin lock, which is an IV catheter kept open with heparin instead of an infusing solution. A heparin lock can be used in many situations instead of a KVO solution.

2. A non-diabetic patient with chest pain being transferred to the coronary care unit. Vital signs are unstable. The patient is hypotensive and has a thready, rapid pulse.
   Answer: 0.9% saline.

3. A diabetic man with polyuria, polydipsia, evidence of modest ECFV depletion, and a blood sugar of 1600 mg/dl. Serum sodium is 155 mEq/L.
   Answer: 0.45% saline. This patient has ECFV depletion and is hypertonic. The 0.45% saline solution will deliver NaCl to the patient to expand the ECFV and free water to correct the severe hypertonicity. Some clinicians would give 0.9% saline first to stabilize the ECFV before starting 0.45% saline.

4. A 35-year-old patient with septic shock.
   Answer: 0.9% saline.

5. A patient with upper GI bleeding. Requires transfusion.
   Answer: 0.9% saline.

6. A diabetic with glucose 1300, sodium 150 mEq/L, BP 60/40, and pulse 120/min.
   Answer: 0.9% saline. The presence of hemodynamic compromise is a higher priority than the hypertonicity. The 0.9% saline should be given first (1–2 liters until the patient is hemodynamically stable), then 0.45% saline to deliver water to correct the hypertonicity.

7. A patient with pulmonary edema. No diabetes.
   Answer: D5W KVO for medications. Alternatively, a heparin lock could be used.

8. An elderly patient. Comatose. Sodium 190 mEq/L. Glucose 100 mg/dl.
   Answer: D5W with frequent determinations of the sodium concentration to avoid overly rapid correction and cerebral edema (see Chapter 4). The patient must be monitored closely for glucosuria. If significant ECFV depletion is present, 0.45 saline could be given first. The most pressing problem in this patient is the life-threatening hypernatremia. D5W delivers 1 liter of water per liter, while 0.45% saline delivers only 500 cc of electrolyte-free water per liter.

9. A 45-year-old patient. Pulmonary and peripheral edema. Sodium concentration 130 mEq/L.
   Answer: The pulmonary and peripheral edema are clinical manifestations of an expanded ECFV. This is caused by an excess of the amount of total body sodium. The patient will need diuretics and sodium restriction to reduce the total body sodium and therefore the size of the ECFV. The patient also has hyponatremia which means that he also has an excess of water relative to sodium in the ECFV. He will therefore require water restriction as well. Because D5W even at a slow rate would give the patient unwanted excess water, I would favor the use of a heparin lock in this case.

10. Write "maintenance" IV orders for a 100 kg man who will be kept NPO (nothing by mouth) for 24 hours for tests. *No renal, cardiac, or liver disease. No recent or future surgery. The patient has no medical condition and is taking no medication listed in **Figs. 3-1** or **3-2** which could cause hyponatremia with hypotonicity.*
    - Amount of water per day: approximately 2500 cc.
    - Amount of sodium per day: approximately 50–100 mEq (for this case, say 100 mEq/day).
    - Amount of potassium per day: approximately 20–60 mEq (for this case, say 60 mEq/day).
    - First, set the rate of the IV by how much water is to be given: 2400 cc/24 hours = 100 cc/hr.
    - Next, set the amount of sodium per liter to be given: 100 mEq/2.4 L = 41.66 mEq/L. The concentration of sodium in D5 0.45% saline is 77 mEq/L. The concentration of sodium in D5W is 0 mEq/L. What if we alternated 1 liter of D5 0.45% saline with 1 liter of D5W? In 24 hours we would deliver 1 liter at 77 mEq/L + 1 liter at 0 mEq/L and 400 cc at 77 mEq/L = 0.4 L = 77 mEq/L = 30.8 mEq. This totals to 77 + 30.8 = 107.8 mEq of sodium, which is close enough.
    - Next, set the concentration of potassium in each IV: 60 mEq/2.4 L = 25 mEq/L. Potassium for IV administration does not generally come in 25

mEq ampules; it comes in 20 mEq and 30 mEq ampules. We could pick either 20 or 30 mEq/L or we could alternate 20 and 30 mEq/L.

- The final IV order might look something like:
  Liter #1: D5 0.45% saline with 30 mEq/L KCl at 100 cc/hour
  Liter #2: D5 W with 20 mEq/L KCl at 100 cc/hour
  Liter #3: D5 0.45% saline with 30 mEq/L KCl at 100 cc/hour
- An IV order is not complete until the monitoring orders are written: Daily weight in the morning. Daily glucose, sodium, potassium, chloride, bicarbonate, blood urea nitrogen (BUN), and creatinine (Cr) in the morning. *A potential complication of hypotonic solutions is severe, acute, life-threatening hyponatremia in a patient who is postoperative or who has an underlying condition capable of causing hyponatremia.* Fluid therapy must be reassessed daily.
- Many clinicians use a more approximate approach to writing IV orders (and so do I). This lengthy example is designed to provide the beginner with a systematic approach to figuring out roughly how much water, sodium, and potassium to give over a given time period. Normally, I would not write "maintenance" orders that require the ward personnel to change the IV solution after each liter. I might just give D5 0.45% saline with 30 mEq/L in this patient for the first day, then D5W with 20 mEq/L KCl the next day. I would check the weight and chemistries each day.

11. Write "maintenance" IV orders for a 50 kg woman who will be kept NPO for 24 hours for tests. *No renal, cardiac, or liver disease. No recent or future surgery. The patient has no medical condition and is taking no medication listed in **Figs. 3-1 or 3-2** which could produce hyponatremia with hypotonicity.*
    - Amount of water per day: approximately 2000–2500 cc (for this case, say 2000 cc).
    - Amount of sodium per day: approximately 50–100 mEq (for this case, say 50 mEq/day).
    - Amount of potassium per day: approximately 20–60 mEq (for this case, say 40 mEq).
    - First, set the rate of the IV by how much water is to be given: 2000 cc/24 hours = 83.33 cc/hr (round to 80 or 85).
    - Next, set the amount of sodium to be given 50 mEq/2.0 L = 25 mEq/L. The concentration of sodium in D5 0.45% saline is 77 mEq/L. The concentration of sodium in D5 water is 0 mEq/L. What if we alternated 1 liter of D5 0.45% saline with 2 liters of D5W? In 24 hours we would deliver 1 liter at 77 mEq/L + 1 liter at 0 mEq/L = 77 mEq. The following day we could give 2 L of D5 water. This would result in 77 mEq sodium given over two days which would be an average of 77/2 = 38.5 mEq per day.

- Next, set the concentration of potassium in each IV: 40 mEq/2.0 L = 20 mEq/L. Potassium for IV administration comes in 20 mEq ampules. No problem.
- The final IV order would look something like:
  Liter #1: D5 0.45% saline with 20 mEq/L KCl at 85 cc/hour
  Liter #2: D5W with 20 mEq/L KCl at 85 cc/hour
  Liter #3: D5 0.45% saline with 20 mEq/L KCl at 85 cc/hour
- Remember that these calculations are approximations. Therapy must always be monitored and reassessed. An IV order is not complete until the monitoring orders are written: Daily weight in the morning. Daily glucose, sodium, potassium, chloride, bicarbonate, blood urea nitrogen (BUN), and creatinine (Cr) in the morning. *A potential complication of hypotonic solutions is severe, acute, life-threatening hyponatremia in a patient who is postoperative or who has an underlying condition capable of causing hyponatremia.*

CHAPTER 3. HYPONATREMIA

A low serum sodium concentration (<135 mEq/L) does not tell us whether total ECF sodium is increased, decreased, or normal. It does tell us that there is an excess of water relative to sodium. *Most cases of hyponatremia are caused by impaired renal water excretion in the presence of continued water intake.* If hyponatremia develops rapidly, there may be severe symptoms caused by brain swelling, such as lethargy, coma, and seizures. If the same degree of hyponatremia develops slowly over several days, there may be no symptoms at all. A patient who is severely symptomatic from hyponatremia needs urgent therapy, while a mildly symptomatic or asymptomatic patient should be treated more gradually.

Causes of Hyponatremia

In most causes of hyponatremia, the blood is hypoosmolal (see **Fig. 3-1**) but two situations in which the ECF is neither hypoosmolal nor hypotonic are discussed first:

- Pseudohyponatremia
- Hyponatremia with hypertonicity

Pseudohyponatremia

Pseudohyponatremia is a very rare situation in which the serum sodium concentration is found to be low but extracellular fluid osmolality and tonicity are *normal*. The low sodium concentration is an *artifact* due to accumulation of other plasma constituents (either triglycerides or protein) in plasma. Pseudohyponatremia occurs in three situations:

- Severe hypertriglyceridemia (triglyceride concentrations in the thousands of mg/dl)
- Severe hyperproteinemia, as may occur in multiple myeloma (plasma protein concentration >10 gm/dl)
- Board Exams (the most common situation, and, honestly, the main reason for this discussion)

FIGURE 3-1. Causes of Hyponatremia

Pseudohyponatremia (rare special case)
  Marked hypertriglyceridemia
  Hyperproteinemia
Hyponatremia with hypertonicity (special case)
  Severe hyperglycemia
  Hypertonic mannitol
Hyponatremia with hypotonicity (requires water intake)
  Renal failure (reduced GFR)
  ECFV depletion (increased reabsorption of water)
  Edematous states (increased reabsorption of water)
  Thiazide diuretics (tubular effect impairing water excretion)
  SIADH: ADH release/effect causing water retention (see **Fig. 3-2**)
  Endocrine: Hypothyroidism or adrenal insufficiency
  Diminished solute intake: "Tea and toast" diet or excessive beer drinking

---

The measured serum osmolality is normal, but the calculated osmolality is low because of the artifactually low serum sodium. Therefore, the osmolal gap is increased. The patient is not symptomatic from the hyponatremia because tonicity is normal. No treatment is required for the low serum sodium concentration. Pseudohyponatremia does not occur when a sodium electrode is used to measure the sodium concentration in an undiluted sample. The sodium electrode technique is now in wide clinical use, so pseudohyponatremia is especially rare nowadays.

Hyponatremia with Hypertonicity

Hyponatremia with *hypertonicity* is another special case of hyponatremia, most often caused by severe hyperglycemia in uncontrolled diabetes mellitus. The sodium is low because of transcellular shifting of water, but both tonicity and measured serum osmolality are very high. Because glucose is an effective osmole, the high glucose concentration causes water movement from the intracellular compartment to the extracellular compartment, thereby reducing the extracellular sodium concentration. Consequently, the sodium concentration decreases, even though the tonicity of the ECFV is increased. The sodium concentration falls by approximately 1.6 mEq/L for every increase of 100 mg/dl in glucose concentration above 100 mg/dl. *To make the diagnosis of hyponatremia with hypertonicity, measured osmolality must be clearly elevated by the hyperglycemia.*

Administration of hypertonic mannitol may also cause hyponatremia with increased tonicity. This is less common than hyperglycemia, but the mechanism is the same: Mannitol causes movement of water from the cellular compartment with subsequent reduction of the sodium concentration. Measured osmolality and tonicity are increased though the measured serum sodium concentration and calculated osmolality are low.

Hyponatremia with Hypotonicity ("True" Hyponatremia)

Hyponatremia with hypotonicity is by far the most common form of hyponatremia and results from impaired renal water excretion in the presence of continued water intake. Hyponatremia with hypotonicity requires two things:

• Impaired renal water excretion
• Continued water intake

Normally, the kidney excretes excess water by producing a large volume of dilute urine. *Finding the reason why the kidney cannot appropriately excrete excess water is the key to diagnosing the cause of hyponatremia.* The impaired renal water excretion may be due to:

• Impaired GFR (renal failure)
• ECFV depletion (often from vomiting with continued ingestion of water)
• Edematous states: congestive heart failure, cirrhosis, and nephrotic syndrome
• Thiazide diuretics
• Syndrome of inappropriate ADH (SIADH) due to a variety of causes (**Fig. 3-2**)
• One of two endocrine abnormalities: hypothyroidism or adrenal insufficiency
• Markedly decreased solute intake combined with high water intake ("tea and toast diet" and excessive beer drinking)

Any of these states that impair water excretion can produce hyponatremia in a patient with a normal serum sodium concentration if sufficient free water is supplied. Therefore, a patient who has one of the conditions listed above is *at risk* for developing hyponatremia if given hypotonic IV fluids or a sudden water load.

Impaired glomerular filtration rate (renal failure)

In order for the kidney to excrete excess water by producing a large volume of dilute urine, there must be an adequate glomerular filtration rate. Obviously, if one cannot filter a water load, then it cannot be excreted! Generally, there needs to be a marked reduction in glomerular filtration rate to around 20% of normal to cause a serious problem with water handling. If there is intake of large amounts of water, however, then less renal impairment would suffice to result in hyponatremia.

ECFV depletion

Although ECFV depletion may arise from many causes, the most common cause associated with hyponatremia is gastric losses from vomiting with concomitant water ingestion (water can be absorbed very rapidly even in the presence of vomiting). Severe depletion of the ECFV also results in ADH release, which contributes to the development of hyponatremia. In ECFV depletion, the proximal tubule is retaining both sodium and water appropriately. The urine sodium concentration is often low (<10 mEq/L) in ECFV depletion, because of *appropriate* renal retention of sodium. The urine volume may also be low (<500 ml/24 hours).

Edematous states

Hyponatremia can occur in decompensated congestive heart failure, cirrhosis, and nephrotic syndrome. As for all of the other causes of hyponatremia with hypotonicity, impaired renal water excretion accompanied by continued water intake is the reason for the hyponatremia. The edema results from a defect in sodium excretion. Hyponatremic patients with edematous states have abnormal renal retention of both sodium (causing ECFV overload and edema) and water (causing hyponatremia). The urine sodium concentration is often low (<10 mEq/L) in edematous states because the kidney is *abnormally* retaining sodium.

Thiazide diuretics

Thiazide diuretics impair renal water excretion by blocking the kidney's ability to produce a dilute urine (Chapter 1). Thiazide diuretics are an important cause of hyponatremia, especially in elderly women. Such hyponatremia may be severe if solute intake is low or water intake is high. *Thiazide diuretics are contraindicated in all patients with hyponatremia, including those with edema-forming states.*

SIADH

What would happen if ADH secretion persisted in spite of a falling serum sodium concentration and continued water intake? Water would continue to be retained, and the serum sodium concentration would continue to decline. This is the basis for the syndrome of inappropriate ADH (SIADH), which originates in various ways (see **Fig. 3-2**):

- Abnormal increase in pituitary ADH secretion
- Ectopic production of tumoral ADH
- ADH-like effect on the collecting tubule by exogenous substances such as medications
- Potentiation of the renal tubular effect of ADH by drugs

Any of these mechanisms may produce the same problem: hyponatremia.

Any serious CNS problem (tumor, infection, or trauma) and many lung problems (especially small cell carcinoma and pulmonary tuberculosis) can cause SIADH. The *postoperative state* is often associated with an increased

FIGURE 3-2. Syndrome of Inappropriate ADH (SIADH)

Central nervous system diseases
  Brain abscess
  Brain tumor
  Meningitis
  Subarachnoid hemorrhage
  Subdural hematoma
  Stroke
  Trauma
Pulmonary diseases
  Bacterial pneumonia
  Acute respiratory failure (ARDS)
  Pulmonary tuberculosis
Neoplasia
  Small cell carcinoma of the lung
  Pancreatic carcinoma
  Carcinoma of the duodenum
Nausea
Postoperative state
Drugs
  Amitriptyline
  Carbamazepine
  Chlorpropamide
  Clofibrate
  Cyclophosphamide
  Haloperidol
  Narcotics
  Nicotine
  Nonsteroidal anti-inflammatory drugs
  Serotonin-reuptake inhibitors
  Thiothixene
  Thioridazine
  Vincristine

release of ADH. A number of commonly used medications may produce this syndrome, either by increasing ADH release, exerting an ADH effect on the kidney, or potentiating the effect of endogenous ADH.

Hypothyroidism and adrenal insufficiency

Both of these conditions are important to consider in cases of undiagnosed hyponatremia, because they represent reversible causes of hyponatremia. The mechanisms for hyponatremia in these conditions are complex.

Diminished solute intake: "tea and toast" diet and excessive beer drinking

The ability of the kidney to defend against hyponatremia depends upon three factors:

- Solute intake and excretion
- The ability of the kidney to excrete water by producing large amounts of dilute urine
- Water intake

Elderly patients have diminished ability to excrete excess water and may develop hyponatremia on a diet low in solutes (i.e., reduced intake of protein and NaCl). The diagnosis may be made with clinical and dietary history and measurement of 24-hour urine osmolal excretion, which is reduced in these cases. The normal range of 24-hour osmolal excretion is about 600–900 mOsm/24 hours. The hyponatremia resolves when the amount of dietary solute is increased by increasing the dietary protein and NaCl and by decreasing water intake. The hyponatremia that occurs in patients with very heavy beer drinking is a similar problem: A large intake of fluid that is low in solute. Beer has considerable carbohydrate, but carbohydrate does not present a large solute load to the kidney because it is ultimately metabolized to CO_2 and water.

A Few Comments about the Patient "at Risk" for Hyponatremia

Any of the states that impair water excretion can produce hyponatremia in a patient with a normal serum sodium concentration if sufficient free water is supplied. Therefore, a patient who has a condition or is taking a drug that impairs water excretion is at risk for developing hyponatremia if given hypotonic IV fluids or a sudden water load. Common situations in which hypotonic fluids can lead to dangerous hyponatremia are ECFV depletion and the postoperative state, but *virtually any of the causes of hyponatremia with hypotonicity listed in* **Fig. 3-1** *and any of the causes of SIADH listed in* **Fig. 3-2** *can place a patient at risk for severe hyponatremia.*

Diagnosis of Hyponatremia

A general approach to the patient with hyponatremia is illustrated in **Fig. 3-3.** The osmolality should be checked: A low measured osmolality indicates hyponatremia with hypotonicity and excludes pseudohyponatremia and hyponatremia with hypertonicity.

History

Perform a careful history to search for causes of ECFV depletion (especially recent vomiting). Ask for a history consistent with congestive heart failure, nephrotic syndrome, chronic renal failure or cirrhosis, ingestion of thiazide diuretics or any of the conditions or medications listed in **Fig. 3-2.** What put this patient at risk for hyponatremia? Ask an elderly patient about dietary protein and salt intake.

FIGURE 3-3. Approach to the Patient with Hyponatremia

First things first: Is the patient severely obtunded or having seizures?
　　　　　　If so, institute emergency treatment while the workup is underway. If the patient is less severely symptomatic (for example, slightly lethargic but easily arousable), and you are not sure what is causing the hyponatremia, then temporary water restriction (<800 ml/24 hours) for a few hours can stop the serum sodium from falling even further while you are making the diagnosis.

Exclude pseudohyponatremia and hyponatremia with increased tonicity.
　　　　　　Check measured plasma osmolality. Is it low and does it correlate with calculated Posm?

Investigate hyponatremia with hypotonicity. Begin by asking two key questions:

(1) Why is renal water excretion impaired?
　　　　　　Renal failure—reduced GFR (*1)
　　　　　　ECFV depletion—increased reabsorption of water (*2)
　　　　　　Edematous states—increased reabsorption of water (*2)
　　　　　　Thiazide diuretics—tubular effect impairing water excretion (*3)
　　　　　　ADH release/effect causing water retention **(Fig. 3.2)**(*4)
　　　　　　Endocrine: Hypothyroidism/ adrenal insufficiency
　　　　　　Diminished solute intake: "Tea and toast" diet and heavy beer drinking

(2) What is the patient's source of excess free water? What are the IV fluids?
　　　　　　Carefully review all oral intake, IV fluids, IV medications (often given with hypotonic fluids). Temporary water restriction for a few hours to 800 ml/24 hours will usually halt a falling serum sodium concentration.

A step-by-step approach to diagnosis: Find the reason for impaired water excretion.

Step 1: Is renal failure present?
　　Check the serum creatinine.

Step 2: Are there signs of ECFV depletion?
• History of nausea, vomiting, or other source of ECFV depletion with water ingestion. Check skin turgor, mucosa, and look for orthostatic fall in blood pressure.
• Is the urine sodium low (<20 mEq/L)? This points to increased sodium reabsorption caused by ECFV depletion. It is important to remember, however, that the edematous states are also associated with a low urine sodium due to abnormal sodium retention by the kidney, but edematous states can easily be distinguished from ECFV depletion on clinical grounds in most cases.

Step 3: Are there signs of ECFV overload?
• Careful history and physical: Search for congestive heart failure (jugular venous distention, pulmonary rales, pleural effusion, ascites, S3 gallop, pretibial edema).
• Cirrhosis (physical findings of chronic liver disease with edema, ascites)
• Nephrotic syndrome (dipstick urine for protein)
• Is the urine sodium low (<20 mEq/L)? This points to increased sodium reabsorption caused by abnormal sodium retention by the kidney. It is important to remember that ECFV depletion is also associated with a low urine sodium caused by the appropriate retention of sodium, but edematous states can easily be distinguished from ECFV depletion on clinical grounds in most cases.

Step 4: Is the patient taking thiazide diuretics?
An important cause of hyponatremia, especially elderly women.

Step 5: Is there a condition or drug capable of producing SIADH (Fig. 3-2)?

Step 6: Is there evidence of thyroid or adrenal insufficiency?
If there is any question as to whether these may be present, the appropriate hormonal assays should be done.

Step 7: Elderly/poor solute intake?
24-hour total solute excretion <600 mOsm/24 hour (This can be measured in a 24-hour urine collection.)

Carefully checking the IV fluids is part of a complete history: If the hyponatremia developed in the hospital, review the intravenous fluids: Hypotonic fluids (D5W, D5 0.45% saline, or 0.45% saline) may have been administered to manage ECFV depletion, for postoperative care, or to a patient otherwise at risk for hyponatremia. Hypotonic fluids should be stopped immediately. Remember that hyponatremia with hypotonicity requires two things:

- Impaired renal water excretion
- *Continued water intake*

Physical Examination

The examiner should carefully search for signs of ECFV depletion (poor skin turgor, dry mucosa, orthostatic fall in blood pressure) or ECFV overload (jugular venous distention, pulmonary rales, pleural effusion, ascites, S3 gallop, pretibial edema). The clinical assessment of the ECFV is relatively easy in the edematous patient: ECFV is obviously increased. The clinical assessment of the ECFV is more difficult when trying to distinguish between normal ECFV and mild depletion of the ECFV. Sometimes a careful trial of 0.9% saline with close monitoring of the sodium concentration and ECFV status can be helpful in such a situation: In ECFV depletion, the serum sodium concentration will usually begin to correct rapidly. Findings consistent with adrenal insufficiency or hypothyroidism should prompt the appropriate hormonal assays.

Laboratory Studies

A low measured serum osmolality excludes pseudohyponatremia or hyponatremia with increased tonicity. The urine osmolality may give an indication of the severity of the impairment in urine dilution. For example, a patient with a urine osmolality of 200 mOsm/L has less impairment of urine dilution than a patient with a urine osmolality of 600 mOsm/L. The urine sodium concentration may be helpful. A low urine sodium concentration (<20 mEq/L) suggests increased proximal reabsorption of sodium secondary to ECFV depletion, congestive heart failure, nephrotic syndrome, or cirrhosis.

Sometimes, despite careful clinical evaluation, it is not clear whether the patient has mild ECFV depletion or SIADH. This is another situation in which the response of the sodium concentration to slow, *careful* administration of 0.9% saline with close monitoring of the sodium concentration and ECFV status may be helpful diagnostically. In ECFV depletion, the serum sodium concentration will usually begin to correct rapidly. In SIADH, the sodium concentration will usually not change much. Nevertheless, because of the possibility of ECFV overload, 0.9% saline administration is not recommended as a *routine* part of the diagnosis of hyponatremia.

Treatment of Hyponatremia

The goals of treatment of hyponatremia are to carefully correct the serum sodium concentration toward normal and to correct any coexisting alterations in the ECFV (see **Fig. 3-4**). Definitive treatment demands finding and treating the specific underlying cause of impaired renal water excretion. Sometimes, diagnostic evaluation and treatment of hyponatremia must proceed simultaneously.

Acute, severely symptomatic hyponatremia is a medical emergency requiring rapid intervention. Treatment of chronic asymptomatic hyponatremia must be approached cautiously, because overly zealous treatment can lead to serious consequences.

Why Symptoms of Hyponatremia Dictate the Urgency of Treatment

The urgency of therapy depends upon the severity of the symptoms of hyponatremia, which in turn depends upon the magnitude of the hyponatremia and the time course of its development. In general, the shorter the time course of development, the more severe the symptoms and therefore the more urgent is the initiation of therapy. This concept is critical to understanding the management of hyponatremia.

When a significant decrease in ECF osmolality occurs rapidly, there is a sudden shift of water to the cellular compartment (water flows down the osmotic gradient). This causes acute cellular swelling, which may result in acute cerebral edema with lethargy, stupor, coma, seizures, and even death. Acute, severely symptomatic hyponatremia is a medical emergency and demands immediate treatment, usually in an intensive care unit setting. Permanent neurological sequelae or death may occur if this condition is not addressed immediately. *The rapidity of development of the hyponatremia is more important than the actual value of the serum sodium concentration.*

What happens if the fall in ECF osmolality occurs over a longer time period, such as several days? The cell can adapt to a slower decrease in the serum osmolality by transporting potassium, sodium, and other solutes out of the interior of the cell. This removal of solute lowers intracellular osmolality and blunts the movement of water into the cell. Brain cell swelling is not as significant as in acute hyponatremia because of this adaptive loss of intracellular solutes. Patients with chronic hyponatremia are generally less symptomatic, and treatment should be less aggressive than in acute hyponatremia. Correction of chronic, asymptomatic hyponatremia needs to be gradual and careful: Rapid correction of extracellular fluid osmolality will lead to rapid shifts of water out of the cells, which may be harmful. In fact, rapid correction or overcorrection of chronic hyponatremia may result in a potentially fatal neurologic condition known as the **osmotic demyelination syndrome (ODS).** Symptoms of the ODS may develop gradually one to several days after the correction of

47

FIGURE 3-4. Treatment of Hyponatremia

Etiology	Treatment	Remarks
Pseudohyponatremia	None	Very rare. Serum TG usually 5000 mg/dl or greater. Protein (multiple myeloma) usually 10 gm/dl or greater.
Hyponatremia with markedly increased tonicity, usually due to profound hyperglycemia.	0.9% saline until hemodynamically stable, then 0.45% saline	This is the *only situation* in which a patient with hyponatremia might be given hypotonic fluids. Measured and calculated Posm must be *severely* elevated due to marked hyperglycemia.
Renal failure	Water restriction.	Restrict water only if hyponatremia is present.
ECFV depletion	0.9% saline.	Usually in setting of vomiting. Most patients need potassium as well as 0.9% saline.
Edematous states	Water restriction for hyponatremia. Sodium restriction and loop diuretics to remove edema.	Restrict water only if hyponatremia is present.
Thiazide diuretics	Stop thiazides. Replace sodium (orally is usually sufficient) and potassium. Correct hypokalemia.	Thiazides should not be given to any patient with hyponatremia.
Increased ADH	Water restriction. Find and treat underlying cause.	If water excretion is severely impaired, may require NaCl tablets to add solute load or demeclocycline to antagonize ADH.
Hypothyroidism	Definitive therapy is to correct hypothyroidism.	
Adrenal insufficiency	Definitive therapy is to correct adrenal failure. 0.9% saline.	Hyponatremia is the most common electrolyte abnormality found in adrenal insufficiency.
"Tea and toast" diet	Increase dietary solute, and decrease water intake if excessive.	Seen mainly in elderly patients.

the serum sodium concentration. These symptoms may include fluctuation in mental status, seizures, swallowing dysfunction, loss of vision, and in severe cases, quadriplegia. ODS is relatively uncommon but may lead to devastating permanent neurological sequelae.

The appropriate treatment of hyponatremia depends upon whether the hyponatremia is symptomatic (symptomatic hyponatremia is typically acute but not always) or asymptomatic (asymptomatic hyponatremia is typically chronic but not always). The risk of permanent neurological impairment due to untreated cerebral swelling and increased intracranial pressure is greatest in acute, severely symptomatic hyponatremia. Acute hyponatremia that is severely symptomatic is a medical emergency that requires rapid intervention. On the other hand, zealous treatment of chronic, asymptomatic or mildly symptomatic hyponatremia could result in permanent neurological impairment secondary to ODS.

Treatment of Chronic Hyponatremia

The treatment of chronic, asymptomatic hyponatremia depends upon the underlying process that caused the hyponatremia. In general, restriction of water to a total (fluid plus dietary foodstuffs) of approximately 800 ml/24 hours will always work temporarily for a few hours until more information is available. Remember: Continued water intake is required in addition to impaired water excretion for hyponatremia to develop. If water is restricted sufficiently, progressive lowering of the serum sodium concentration will stabilize, regardless of the specific cause of hyponatremia. Water restriction is *not* appropriate treatment for hyponatremic patients with ECFV depletion, thiazide-induced hyponatremia, or hypothyroidism or adrenal insufficiency, but water restriction can serve as a *temporizing measure* for a few hours in cases of uncertain diagnosis. Therefore, if it is necessary to "put the brakes on" in a case of a rapidly falling serum sodium of unclear etiology, water restriction will work temporarily until more information is available and appropriate specific therapy is begun.

The primary way an anephric dialysis patient develops hyponatremia is by drinking water in excess of insensible loss. The treatment of choice for hyponatremia associated with **renal failure** is water restriction. Emergency dialysis can correct acute, symptomatic hyponatremia due to a large, sudden water load in a patient who is anuric.

First-line therapy of hyponatremia associated with the **edema-forming states** is also water restriction. Water restriction treats the *water control* problem (the cause of the hyponatremia). Loop diuretics and sodium restriction treat the *sodium-overload* problem (the cause of the edema). Water restriction is *not* necessary unless there is hyponatremia. It is also important to remember that thiazide diuretics are contraindicated in all patients with hyponatremia, even those with edema-forming states. Thiazides impair the ability of the kidney to produce a dilute urine and can worsen the hyponatremia.

It is important to re-emphasize that in the absence of hyponatremia there is no indication to restrict water. If there is edema (ECF sodium overload) alone, restrict only sodium. If there is hyponatremia and edema, restrict both water and sodium.

The treatment of **ECFV depletion** is volume replacement with 0.9% saline. Generally, the kidney is appropriately retaining sodium, so the urine sodium is typically very low in this condition (<10 mEq/L). Both the hyponatremia and depleted ECFV respond to replacement of ECFV with isotonic saline solution. This results in normalization of the sodium concentration and restoration of ECFV. One needs to be careful because overly rapid correction may lead to the osmotic demyelination syndrome. This is especially true in patients who are asymptomatic or mildly symptomatic whose hyponatremia has developed slowly over several days. Once the ECFV is restored to normal with 0.9% saline, these patients will quickly restore their sodium concentration to the normal range. It is imperative to check the sodium concentration frequently in these patients during therapy.

The treatment of hyponatremia secondary to **thiazide diuretics** is to stop the diuretics and liberalize sodium intake. Potassium depletion is often present as well, and should be corrected. Such patients should not be restarted on thiazides because they are at risk for repeated episodes of dangerous hyponatremia.

For cases of **SIADH**, water restriction is always helpful until the underlying condition is identified and specific therapy is begun. Any medications known to impair water excretion should be withheld (**Figs. 3-1** and **3-2**), and a careful history and physical examination performed. Treatment depends upon whether or not symptoms are present (acute versus chronic hyponatremia). In cases of severe impairment of water excretion, increasing solute intake (NaCl tablets) or administration of agents which counteract the action of ADH, such as demeclocycline, may be required. Definitive therapy is removal of the cause of SIADH.

Recently, a new class of drugs, the vasopressin (ADH) antagonists have been developed, which block the action of ADH at the level of the collecting tubule. By blocking the effect of ADH, these agents can induce water excretion and thereby can produce an increase in the serum sodium concentration. When available, these agents may be clinically useful in treating patients with SIADH from a variety of causes. Although I have classified the hyponatremia associated with the edematous disorders, congestive heart failure and cirrhosis, as being caused by increased proximal water reabsorption leading to decreased water excretion, many of these patients also have abnormally increased ADH contributing to the hyponatremia. Therefore, patients with congestive heart failure or cirrhosis could also benefit from the ADH antagonists. The ADH antagonists are currently investigational.

Treatment of Acute, Symptomatic Hyponatremia

Patients with acute, severely symptomatic (stupor, coma, seizures) hyponatremia require closely monitored emergency care. Water restriction, although not sufficient to correct the hyponatremia quickly enough to relieve symptoms of brain swelling, will prevent the hyponatremia from worsening. In general, the indications for emergency treatment with hypertonic saline are manifestations of significant CNS symptoms, such as severely depressed mental status, seizures, or other evidence of increased intracranial pressure. Acute, severe hyponatremia may result from any of the causes of hyponatremia, but the most frequent causes are:

- Hypotonic fluids (e.g. 0.45% saline, D5 0.45% saline, D5W) administered postoperatively or in the setting of ECFV depletion
- Syndrome of inappropriate ADH in the presence of acute excess water ingestion
- Primary polydipsia leading to acute water intoxication in the psychiatric patient. For hyponatremia to develop, there generally must be an impaired ability to excrete water *in addition* to increased water ingestion.
- ECFV depletion (often secondary to protracted vomiting with continued water ingestion). Generally this form of hyponatremia will respond to volume expansion with 0.9% NaCl and should not require hypertonic saline.
- Intravenous cyclophosphamide (increased ADH effect)

How to use 3% saline

There are important safety guidelines for the cautious use of hypertonic saline. The major risk of using 3% saline is too rapid correction or overcorrection of hyponatremia resulting in ODS. It is important to individualize therapy in each patient, and to make adjustments as the clinical picture changes. The following are general safety guidelines:

- Treat only *severely* symptomatic patients. Asymptomatic or mildly symptomatic patients should not generally be treated with 3% saline.
- In general, correct the sodium concentration no faster than 1 mEq/L per hour initially until achieving 6–8 mEq/L increase, then 0.5 mEq/L (or less) per hour.
- An increase of the sodium concentration of 6–8 mEq/L should be enough to reduce symptoms acutely. Once symptoms improve, the 3% saline should be slowed/stopped.
- Permit no more than 10–12 mEq/L increase in sodium concentration in the first 24 hours.
- 50–100 ml/hr of 3% saline initially is generally safe for a brief period in a severely symptomatic "average" sized person, until appropriate calculations can be done to determine a more precise rate of administration.
- Be very careful in women and in chronically ill patients, such as alcoholics, cancer patients, and patients who have had a recent cardiac arrest because these patients are at risk for ODS.
- Slow/stop the infusion as soon as symptoms improve. *The object is not to correct the serum sodium value per se but to alleviate cerebral edema.*

• Watch for congestive heart failure. Some clinicians give a loop diuretic along with the 3% saline to avoid ECFV overload and to increase water excretion. **A word of caution:** Loop diuretics may make the sodium concentration rise faster than predicted, so the serum sodium concentration must be checked very frequently.

Once the decision is made to use hypertonic saline, the infusion rate of 3% saline may be calculated as follows:

1) Calculate the "sodium deficit."

Calculate the *amount of sodium in mEq* that, if it all remained in the extracellular fluid compartment, would raise the serum sodium concentration to the desired level. Hyponatremia is *not* the result of a deficit of sodium, as the term "sodium deficit" suggests. This is a confusing bit of terminology, but it is useful to help us calculate the *amount of sodium in mEq* to give as 3% saline to raise the sodium concentration by a desired amount.

To do this, first set the desired serum sodium concentration value using the parameters for safe correction discussed above, and the *time* (t, in hours) at which you want the sodium to be its desired value:

• Na^+ (mEq given as 3%) = ([$Na^+_{(desired)}$] − [$Na^+_{(measured)}$]) × Estimated TBW
• This gives the *amount* of sodium in mEq to be given as 3% saline over time t.

2) Set the rate of the infusion.

• There are 513 mEq of sodium in a liter of 3% saline. Therefore, to obtain the volume of 3% saline to give over time t, divide the number of mEq of sodium to be given (which was determined in step 1) by 513 mEq/L.
• Then give this volume of 3% saline over time t.

In general, adequate administration of hypertonic saline will cause a temporary relief of cerebral edema by translocating water out of brain cells. A loop diuretic can eliminate excess water from the body. Therefore, a loop diuretic is sometimes administered with 3% saline in the setting of acute, symptomatic hyponatremia. Loop diuretics may make the sodium concentration rise faster than predicted, however, so the serum sodium concentration must be checked very frequently.

A word of caution: Overly rapid correction can result from an overestimation of TBW just as well as from a miscalculation of the "sodium deficit." Be careful to use .5 × body weight (kg) in women rather than the more widely quoted .6 × body weight (kg), which is used in men. Also note that elderly patients have a decreased TBW. A chronically ill hospitalized patient may have an even lower TBW than predicted based upon age and gender. The key is frequent determination of the serum sodium concentration to monitor therapy.

Correcting acute hyponatremia with 3% saline is an emergency procedure and is best carried out under close observation with frequent determinations of serum sodium to monitor therapy. I strongly favor ICU monitoring for this; I generally calculate the amount of 3% saline to be given for a 4 hour period initially and reassess the sodium concentration during therapy and at the end of this period. I check the sodium concentration every 1–2 hours to carefully monitor therapy.

The purpose of using 3% saline in cases of severe acute, symptomatic hyponatremia is not to correct the sodium concentration per se, but to temporarily relieve cerebral edema and to prevent neurologic sequelae. I view the use of 3% as trying to "let off a little pressure" by the hypertonic effect of the 3% saline, which removes water from the swollen brain. This can generally be accomplished by correcting the sodium concentration by only 6-8 mEq/L initially. I generally do not infuse 3% saline for more than a total of 6–8 hours. Once the sodium is corrected by 6–8 mEq/L, I generally stop the 3% saline and initiate more conservative measures. *It is never necessary to use 3% saline to correct the sodium to even close to the normal range.* Overcorrection is a serious mistake that could cause ODS. I view 3% saline as a potentially dangerous *drug* to be used only in the early treatment of severely symptomatic cases. The problems with ODS seem to occur well *after* the cerebral edema is corrected, so once the patient is out of danger from cerebral edema, I *stop* the hypertonic saline and employ the appropriate measures discussed in the section on chronic hyponatremia.

It is important to individualize therapy for each patient, depending upon the clinical picture and the rate of response of the serum sodium concentration to 3% saline infusion. For example, a patient who is having continuous seizures may require an initial rate of correction of the serum sodium concentration faster than the recommended 1 mEq/L per hour for an hour or two, depending on the patient's clinical response and the rate of rise of the serum sodium concentration.

Exercises

1. General approach to hyponatremia: A colleague calls you on the phone for advice on how to approach a patient with a serum sodium concentration of 120 mEq/L. The patient is asymptomatic. The other data are: chloride 80 mEq/L, potassium 4.5 mEq/L, bicarbonate 24 mEq/L, BUN 14 mg/dl, glucose 90 mg/dl. What further questions do you ask?

   Answer:
   1. What is the measured osmolality? It is 255 mOsm/L (normal range: 280–295 mOsm/L). Therefore, you are dealing with hyponatremia with hypotonicity. The osmolal gap is 5 mOsm/L.
   2. Is there renal failure? The creatinine is 1.0 mg/dl (normal), which rules out renal failure.
   3. Is there evidence of an abnormally increased or decreased ECFV? Your colleague must search carefully for an edematous state or for evidence of ECFV depletion. There is none. The urine sodium concentration may sometimes be helpful in hyponatremia: ECFV

53

depletion and edematous disorders may have urine sodium values <20 mEq/L, indicating renal sodium retention. You ask for the urine sodium concentration. It is 65 mEq/L, which is further evidence against ECFV depletion or an edematous disorder.

4. Is the patient taking thiazides? Hyponatremia may result from thiazide diuretics given to treat hypertension. The patient denies taking thiazides.

5. Is there evidence of a disorder or is the patient taking a medication capable of causing SIADH? (refer to **Fig. 3-2.**) Any of the disorders or drugs listed could be causing the hyponatremia.

6. Is there evidence of adrenal failure or hypothyroidism? If in doubt, your colleague may order the appropriate assays.

7. You also ask about the amount of intake of water and solute. If the patient were hospitalized, you would immediately check to see what IV fluids the patient is receiving.

Of course, a phone conversation is no substitute for a careful clinical evaluation, but these are some key questions to ask when evaluating a patient with hyponatremia.

2. A 30-year-old man with a history of high blood lipids has the following laboratory values: sodium 125 mEq/L, measured osmolality 270 mOsm/L, triglycerides 1000 mg/dl, total protein 8.5 gm/dl. The blood sample is lipemic. Is this a case of pseudohyponatremia?

Answer: No. The measured osmolality is low. This is hyponatremia with hypotonicity. The sodium concentration falls by about 1 mEq/L for every 500 mg/dl increase in triglyceride concentration. For pseudohyponatremia to develop, the degree of hypertriglyceridemia must be severe: Triglycerides must increase by about 5000 mg/dl for a 10 mEq/L drop in the serum sodium concentration. A triglyceride concentration of 1000 mg/dl would result in a fall in the sodium concentration of only about 2 mEq/L. On the other hand, blood becomes lipemic when the triglyceride concentration reaches 500 mg/dl: thus, lipemic blood does not confirm the diagnosis of pseudohyponatremia although the absence of lipemia rules out hypertriglyceridemia-related pseudohyponatremia. Similarly, for pseudohyponatremia to be present in hyperproteinemic states such as multiple myeloma, the protein concentration must be increased by roughly 0.25 gm/dl above 8 gm/dl for every 1 mEq/L fall in the serum sodium concentration. So the protein concentration would need to be 10.5 gm/dl for the sodium concentration to fall by 10 mEq/L. In a nutshell: If the blood of the patient is not lipemic and the protein concentration in plasma is less than 10 gm/dl, then pseudohyponatremia is not present.

Another point is that the measured osmolality is normal and the calculated osmolality is low in cases of pseudohyponatremia. This results in an increased osmolal gap. The osmolality of 270 mOsm/L in this patient is low and indicates one of the causes of hyponatremia with hypotonicity.

[margin note: Pseudohyponatremia → must have normal osmolarity]

54

3. A patient presents with 3 days of nausea, vomiting, polyuria and polydipsia. On physical examination, there is poor skin turgor and a blood pressure drop from 136/86 to 92/52 on standing, with a pulse increase from 96 to 128. Laboratory studies show: glucose 360 mg/dl, sodium 120 mEq/L, and blood urea nitrogen 28 mg/dl. The measured osmolality is 270 mOsm/L. What is the cause of the hyponatremia in this patient, and what would the serum sodium concentration be once it has been "corrected" for the elevated glucose? What fluid would you use?

Answer: The hyponatremia is most likely caused by ECFV depletion, caused by protracted nausea and vomiting, with continued intake of water. Serum osmolality may be calculated (Chapter 1):

$$OSM_{(calc)} = 2(120) + 360/18 + 28/2.8 = 270 \text{ mOsm/L}$$

and is consistent with the measured osmolality. The patient is ECFV depleted and hypotonic. The glucose is contributing only $360/18 = 20$ mOsm/L to the osmolality. When the hyperglycemia corrects, the serum sodium concentration will increase. The correction factor for the serum sodium concentration (that is, to estimate what the serum sodium will increase to once all the glucose is pushed back into cells where it belongs) is 1.6 mEq/L sodium for every 100 mg/dl increase in glucose concentration above 100 mg/dl. The correction factor would be roughly $3 \times 1.6 = 4.8$ mEq/L (rounding the glucose of 360 mg/dl to 400 mg/dl), and the corrected serum sodium concentration would be $120 + 4.8 = 125$ mEq/L. This is roughly the serum sodium that would be expected if the glucose concentration were corrected to 100 mg/dl This patient has ECFV depletion and hyponatremia with hypotonicity. The IV solution of choice is 0.9% saline, which will correct the ECFV depletion and the hyponatremia.

Most uncontrolled diabetics with hyperglycemia are treated with 0.9% saline. This is by far the most common clinical situation.

4. A patient presents with polyuria and polydipsia and altered mental status. On physical examination, the patient is poorly responsive. There is poor skin turgor and a blood pressure drop from 100/84 to 80/62 on sitting, with a pulse increase from 92 to 128. Laboratory studies show: glucose 2100 mg/dl, sodium 130 mEq/L, and blood urea nitrogen 40 mg/dl. The measured osmolality is 395 mOsm/L. What is the cause of the hyponatremia in this patient, and what is the serum sodium once it has been "corrected" for the elevated glucose? What fluid would you give?

Answer: This is a special case of hyponatremia caused by the *massive* elevation of the glucose concentration: Hyponatremia with severe hypertonicity. The patient is obtunded because of the severe hypertonicity. Serum osmolality may be calculated (Chapter 1):

$$OSM_{(calc)} = 2(130) + 2100/18 + 40/2.8 = 391 \text{ mOsm/L}$$

The calculated value of 391 mOsm/L is consistent with the measured osmolality of 395, confirming glucose as the solute responsible for the increased osmolality. The glucose is contributing $2100/18 = 111$ mOsm/L to the osmolality. The extracellular fluid is *hypertonic* and the measured osmolality is markedly increased. There is actually a *deficit* of water relative to solutes. When the hyperglycemia corrects, the serum sodium concentration will increase. The correction factor for the serum sodium concentration (that is, to estimate what the serum sodium will increase to once all the glucose is pushed back into cells where it belongs) is 1.6 mEq/L sodium for every 100 mg/dl increase in glucose concentration above 100 mg/dl. The correction to the serum sodium concentration would be $20 \times 1.6 = 32$ mEq/L for a glucose of 2100 mg/dl, and the corrected serum sodium concentration would be $130 + 32 = 162$ mEq/L, which is the serum sodium that would be expected if the glucose concentration were corrected to 100 mg/dl. This corrected value for the serum sodium concentration (162 mEq/L in this case) should be considered when deciding upon which IV fluid to use.

Many clinicians would give a liter of 0.9% saline first for rapid correction of the severe volume depletion, followed by hypotonic 0.45% saline to give free water as well as sodium (volume) replacement. This is usually appropriate, especially when there is significant hemodynamic compromise. The intravenous fluid best suited for subsequent definitive treatment of this situation (hypertonicity and volume depletion) is 0.45% saline. Why? Because the patient is *markedly hypertonic*, the IV fluid should be hypotonic relative to plasma to provide the water the patient needs. The patient is also volume depleted as indicated by the poor skin turgor and orthostatic blood pressure drop, so the patient will need sodium as well. The solution that provides both free water to correct the hypertonicity as well as sodium to correct the depleted ECFV is 0.45% saline. D5W would give free water, but the dextrose in this solution would aggravate the hyperglycemia. D5W contains no sodium, which is needed to correct the volume depletion. Hyponatremia with severe hypertonicity is a special situation. In most other circumstances, hypotonic fluids should not be given to a patient with ECFV depletion because hypotonic fluids deliver free water to a patient who will have difficulty excreting it and could therefore result in dangerous hyponatremia.

5. What if a patient had a sodium concentration of 120 mEq/L, a BUN of 28 mg/dl, a measured osmolality of 275, and a glucose of 360 mg/dl? Would you give 0.45% saline?
   Answer: No. This is *not* a case of hyponatremia with hypertonicity. In this case, the calculated osmolality is:

$$OSM_{(calc)} = 2(120) + 360/18 + 28/2.8 = 270 \text{ mOsm/L}$$

This is hyponatremia with hypotonicity. *Hypotonic fluids are absolutely contraindicated in hyponatremia with hypotonicity.* Just because the glucose is elevated does not mean the patient has hyponatremia with hypertonicity. This patient needs further evaluation to determine the cause of the hyponatremia. The glucose is not high enough to make the patient hypertonic, nor to significantly lower the serum sodium. The glucose is raising the osmolality by only $360/18 = 20$ mOsm/L. To make the diagnosis of hyponatremia with hypertonicity requires a much higher glucose concentration. This case illustrates the importance of checking both the calculated and measured serum osmolalities in patients with hyponatremia. Hypotonic fluids should never be given to a patient with hyponatremia except in the special case of *severe hypertonicity*.

6. What if a patient had a sodium concentration of 120 mEq/L, a BUN of 280 mg/dl, a glucose of 360 mg/dl, and a measured osmolality of 365? Would you give 0.45% saline?
   Answer: No. This is *not* a case of hyponatremia with hypertonicity, even though the calculated osmolality is:

   $$OSM_{(calc)} = 2(120) + 360/18 + 280/2.8 = 360 \text{ mOsm/L}$$

   In this case, the osmolality is elevated mainly by urea which is an ineffective osmole. Urea is contributing $280/2.8 = 100$ mOsm/L to the osmolality, but the patient is not hypertonic because urea is not an effective osmole and does not contribute to tonicity. Glucose is an effective osmole, but is raising the osmolality by only $360/18 = 20$ mOsm/L. The glucose is not high enough to make the patient hypertonic, and is not the cause of the low serum sodium concentration. This is a case of hyponatremia with hypotonicity, even though the osmolality is markedly increased. Hypotonic fluids are absolutely contraindicated in hyponatremia with hypotonicity. This patient needs further evaluation to determine the cause of the hyponatremia.

7. A neurosurgical resident shows you the following laboratory results: sodium 130 mEq/L, measured osmolality 330 mOsm/L, glucose 180 mg/dl, BUN 28. The patient is receiving hypertonic mannitol for cerebral edema. How would you suspect that mannitol is contributing to a patient's hyponatremia?
   Answer: The measured osmolality is very high, and the calculated osmolality is low, resulting in a large increase in the osmolal gap.

   $$\text{Calculated osmolality} = 2(130) + 180/18 + 28/2.8 = 280 \text{ mOsm/L}$$

   The osmolal gap is the difference between the measured and calculated osmolal gaps and is normally no greater that 10 mOsm/L:

   $$\text{Osmolal Gap} = 330 \text{ mOsm/L} - 280 \text{ mOsm/L} = 50 \text{ mOsm/L}.$$

   The osmolal gap is elevated by mannitol. This leads to hyponatremia with hypertonicity.

8. Use and interpretation of the osmolal gap: A patient comes to the emergency room staggering and smelling like beer. The sodium is 140 mg/dl, glucose 180 mg/dl and BUN 28 mg/dl. The measured osmolality is 330. What is your diagnosis and how do you confirm it?

Answer: The osmolal gap is:

$$330-[2 \times (140) + 180/18 + 28/2.8] = 330-300 = 30$$

This patient is drunk. The most likely substance causing an increase of the osmolal gap is ethanol, which should be confirmed by ordering an ethanol level. Other substances that can elevate the osmolal gap are ethylene glycol, methanol, isopropanol, mannitol and sorbitol.

9. Mr. Jones is a 78-year-old retired minister who has end-stage renal disease secondary to glomerulonephritis and zero urine volume/24 hours. He has been a very compliant dialysis patient, but today he comes to dialysis weighing an extra 5 kg and his serum sodium has fallen from 136 mEq/L to 124 mEq/L. He repeatedly maintains that he has not taken in more than the prescribed amount of fluid (1.5 L/day) between his dialysis sessions. What is the cause of the hyponatremia?
Answer: He has taken in too much fluid! There is no other possibility in a man who has no GFR. Generally, the solution to such a problem turns out to be a matter of more patient education, because many patients often do not consider such items as gelatin, soup and broth, or milk added to cereal in the calculation of "water" intake.

10. A 54-year-old man presents with increasing shortness of breath, fatigue, paroxysmal nocturnal dyspnea, and marked edema. His physical examination reveals jugular venous distention, rales, and a third heart sound (S3). His chest X-ray shows bilateral pleural effusions, cardiomegaly, and interstitial pulmonary infiltrates. His serum sodium is 125 mEq/L. The urine sodium concentration is 5 mEq/L (<20 mEq/L indicates renal retention of sodium). Measured osmolality = 270 mOsm/L. Why is the serum sodium concentration low? What therapy would you give?
Answer: It is apparent that this patient has severe congestive heart failure. The patient has gross clinical and radiographic evidence of ECFV overload. The urine sodium is low (<20 mEq/L) which indicates abnormal renal retention of sodium. The abnormal renal sodium retention has led to a striking excess in total body sodium. Note that the ECFV, which tells us about sodium balance, is assessed by clinical criteria: Edema, jugular venous distention, and congestive symptoms along with the chest X-ray showing volume overload. Why is the sodium concentration low? The reason for the hyponatremia is that the patient has an impairment of renal water excretion. The low serum sodium tells us absolutely nothing about total body sodium.

In a systematic way, assess the ECFV (determined by total body sodium) first, then the sodium concentration (which tells us about water regulation):
- Assess the ECFV: Edema, JVD, congestive symptoms, rales, S3, and chest X-ray showing pulmonary edema all indicate a marked excess of total body sodium. The treatment is *loop* diuretics and sodium restriction.

- Hyponatremia: The serum sodium of 125 mEq/L means water excess relative to sodium, which is due to impaired renal water excretion in the presence of ongoing water intake. The treatment is water restriction. *If the serum sodium were normal, there would be no indication for water restriction.*

It is better to consider sequentially the size of the ECFV, then the sodium concentration. Look for clinical indices of abnormal ECFV: Overload produces edema, JVD, rales, S3, weight gain, CVP elevation, and pulmonary congestion on CXR. Depletion produces hypotension, tachycardia, orthostatic fall in blood pressure with increase in pulse, dry mucous membranes, poor skin turgor, and low CVP. After clinically assessing the ECFV (which is determined by total body sodium), look at the serum sodium concentration (which tells you about water regulation).

11. A 59-year-old woman presents with congestive heart failure and a serum sodium of 138 mEq/L. Do you restrict water, sodium, or both?
    Answer: Restrict only sodium. *In the absence of hyponatremia there is no indication to restrict water* in congestive heart failure, nephrotic syndrome, or cirrhosis with edema.

12. A 26-year-old woman with a history of peptic ulcer disease presents with nausea, vomiting, and abdominal pain. On examination, her mucous membranes are dry, her skin turgor is poor, and her blood pressure falls from 120/80 to 85/60 mm Hg upon standing from the supine position, with a concomitant rise in pulse from 95 to 120. Her serum sodium is 125 mEq/L. The urine sodium concentration is 5 mEq/L (<20 mEq/L indicates renal retention of sodium). Measured osmolality = 270 mOsm/L. What is happening? What therapy do you prescribe?
    Answer: This patient suffers from marked ECFV depletion. Her total body sodium is severely decreased. The urine sodium is low due to *appropriate* renal conservation of sodium. The patient is also retaining water because water is being avidly reabsorbed by the proximal tubule and therefore cannot be excreted. The reason for the hyponatremia is that the patient has impaired water excretion in the presence of continued water intake. In a systematic way, assess the ECFV (determined by total body sodium) first, then the sodium concentration (which tells us about water regulation):
    - Assessment of ECFV: The dry mucous membranes, decreased skin turgor, and orthostatic fall in blood pressure all support a diagnosis of ECFV depletion and therefore a marked depletion of total body sodium. The treatment is 0.9% saline. *Hypotonic fluids should not be used because they deliver electrolyte-free water and can worsen the hyponatremia.*
    - Assessment of the sodium concentration (which tells us about water regulation): The serum sodium of 125 mEq/L means there is an excess of water relative to sodium. In this case the sodium concentration will correct with restoration of ECFV with 0.9% saline.

13. A 72-year-old man who is a heavy cigarette smoker presents with increasing cough and one episode of hemoptysis. Other than nicotine stains on his right index and middle fingers, his physical examination is normal. Chest X-ray reveals a 4 cm right lung mass. His serum sodium is 125 mEq/L, potassium 4.2 mEq/L, creatinine 1.1 mg/dl. Measured osmolality = 270 mOsm/L. The urine sodium is 45 mEq/L. His mucous membranes are moist, skin turgor is normal, and he does not have an orthostatic fall in blood pressure. He takes no medications. TSH is normal. What is causing the hyponatremia?

Answer: This patient does not exhibit either an obvious excess or a depletion of ECFV (total body sodium). The low serum sodium tells us nothing about the total body sodium status of the patient, but rather indicates that there is an excess of water relative to sodium, which is due to an impairment of renal water excretion. Systematically:

1. What is the measured osmolality? It is 270 mOsm/L (normal range: 280–295 mOsm/L). Therefore, you are dealing with hyponatremia with hypotonicity.
2. Is there renal failure? The creatinine is 1.1 mg/dl (normal), which rules out renal failure.
3. Is there evidence of an abnormally increased or decreased ECFV? There is none. The urine sodium concentration is 45 mEq/L, which is further evidence against ECFV depletion or an edematous disorder.
4. Is the patient taking thiazides? No.
5. Is there evidence of a disorder or is the patient taking a medication capable of causing SIADH? (refer to **Fig. 3-2.**) We suspect the lung mass may be causing SIADH. Any of the other disorders or drugs listed could be causing or worsening hyponatremia as well.
6. Is there evidence of adrenal failure or hypothyroidism? TSH is normal. Adrenal failure is a very remote possibility.

14. An elderly woman presents with a sodium concentration of 125 mEq/L and a potassium concentration of 3.4 mEq/L. She is alert but complains of difficulty with her memory. The measured osmolality is 270 mOsm/L. There is a history of hypertension, but the patient does not remember the name of her medication. She is asymptomatic except for some trouble with recent memory. What is your diagnosis?

Answer: Use the same routine as in the previous exercise. In this woman, consider thiazide-induced hyponatremia. She has her husband bring her medicine to you: it is hydrochlorothiazide. The drug should be stopped and dietary sodium and potassium liberalized. The woman should no longer be treated with thiazides, because this situation may recur. An elderly patient may also have some inherent difficulty with water excretion as well as a decreased solute intake contributing to hyponatremia (see exercise 16).

15. A 33-year-old 60 kg advertising executive is recovering from a resection of a tubo-ovarian abscess. The intravenous fluid she is receiving is D5

0.45% saline. Thirty six hours after surgery she is noted to be poorly arousable and suffers a generalized seizure. Her serum sodium concentration has fallen from 136 mEq/L before surgery to 116 mEq/L. Measured osmolality = 258 mOsm/L. What has happened? What do you do?

Answer: Postoperative patients often have an impaired ability to excrete a water load, probably because of stimulated ADH release. *Hypotonic fluids such as D5W, D5 0.45% saline, and 0.45% saline should not generally be used postoperatively or in the setting of volume depletion, because they deliver free water to a patient who will have difficulty excreting it, potentially leading to dangerous hyponatremia.* Isotonic fluids are best used in these situations. This patient has retained the water given to her as D5 0.45% saline. You should ask for the IV to be stopped and for 3% saline to be started. You can safely begin the 3% saline infusion with a rate of 50-100 ml/ hour for a brief time until you can calculate a more precise rate of administration. Remember: There are important safety guidelines for the cautious use of hypertonic saline that should be thoroughly reviewed prior to each use of 3% saline. The major risk of using 3% saline is too rapid correction or overcorrection of hyponatremia resulting in ODS.

To calculate the amount of 3% to give, specify a *time* at which you would like to remeasure the serum sodium. Let's say 4 hours, because we will need to check the sodium very frequently. At 4 hours, then, what would you like the serum sodium to be? Carefully reviewing the safety guidelines for rapid correction of hyponatremia with 3% saline, you might want the sodium concentration in this patient to be perhaps 120 mEq/L. Now use *the equation:*

- Na^+ (amount in mEq to be given as 3%) = ([Serum $Na^+_{(desired)}$] − [Serum $Na^+_{(measured)}$]) × Estimated total body water (TBW)
- This gives the *amount of sodium in mEq* to be given as 3% saline over time t.
- Na^+ (mEq) = (120-116) × (0.5 × 60 kg)
- Na^+ (mEq) = 120 mEq

So 120 mEq of sodium is to be given as 3% saline over the next 4 hours. Because 3% saline has 513 mEq sodium/L, the volume of 3% saline would be: 120/513 = 0.234 L = 234 ml over 4 hours. The sodium concentration is rechecked every 1-2 hours to monitor therapy. Administration of 3% saline is mainly a temporizing measure to relieve cerebral edema. Sometimes a loop diuretic may be given along with the 3% saline. The loop diuretic has the effect of eliminating water from the body and therefore constitutes a more definitive treatment of the acute hyponatremia. Loop diuretics may make the sodium concentration rise faster than predicted, however, so the serum sodium concentration must be checked very frequently.

16. The effect of solute intake on the ability to excrete water: The daily urine volume normally ranges from as high as 18 L to as little as .5 L. The

normal daily obligatory solute excretion is approximately from 600 to 900 mOsm and consists of urea, electrolytes (sodium, potassium, and their attendant anions), and waste products originating from the diet. The normal kidney is able to dilute the urine to as little as 50 mOsm/L or to concentrate the urine to as high as 1200 mOsm/L. Therefore, the urine volume in 24 hours could be as high as 900 mosm/50 mOsm/L = 18 L in a maximally dilute urine and as low as 600 mOsm /1200 mOsm/L = 0.5 L in a highly concentrated urine.

The prevention of hyponatremia depends upon the balance of three factors:
• Solute intake
• The ability of the kidney to dilute the urine maximally
• Water intake

The following is intended as a semiquantitative discussion. The following calculations are *approximations*, but serve to illustrate the interactions of these three factors. More precise and rigorous calculations based upon electrolyte free water excretion are beyond the scope of this text.

Solute intake and the ability of the kidney to produce a dilute urine will determine the maximum amount of water intake allowable before hyponatremia develops. The ability to make a dilute urine declines with aging in some persons so that the minimum urine osmolality achievable is 100-150 mOsm/L instead of 50 mOsm/L. Therefore, assuming a maximally dilute urine of 150 mOsm/L, the maximum amount of water ingested in a 24-hour period by an elderly person without lowering the serum sodium is about 900 mOsm/150 mOsm/L = 6 L + insensible loss (1/2 L) = 6 1/2 L. Water ingested in excess of 6 1/2L would dilute the sodium concentration, but a water intake less than 6 1/2 L should not lead to hyponatremia. Fortunately, few people will drink 6 1/2 L!

Now, consider the example of the elderly patient ingesting a low solute diet. In this case, the person has a decreased solute load to excrete because the diet is low in protein and electrolytes. The solute load may be as low as 300 mOsm/day. What would the maximum water intake be such that the sodium concentration would not fall? In other words, how much water can this person drink without developing hyponatremia?
Answer: 300 mOsm/150 mOsm/L + 1/2 L (insensible) = 2 1/2 L! Consequently, an elderly person with mild impairment of diluting capacity and with a poor solute intake will develop hyponatremia if drinking more than 2 1/2 L/day! This is the mechanism behind the hyponatremia of the "tea and toast" diet. Note the relationship among the 3 factors: If the renal solute excretion were increased from 300 to 600 mOsm/day, the water intake could be increased to 600/150 + 1/2 L = 4 1/2 L/day and the patient would be much less likely to develop hyponatremia.

17. Using the approximate calculations described in the previous exercise, what would the allowable water intake be without risk of hyponatremia

in a man with SIADH, a fixed urine osmolality of 400, and a dietary solute load of 400 mOsm/day?

Answer: 400 mOsm/400 mOsm/L + 1/2 L = 1 1/2 Liters. To increase this allowable amount of water, we could either increase the solute load, perhaps by giving NaCl tablets; or somehow decrease the urine osmolality, perhaps by giving demeclocycline to counteract the effects of ADH on the collecting tubule. If the solute load were increased to 800 mOsm/day, for example, then the allowable water intake would be roughly: 800 mOsm/400 mOsm/L + 1/2 L = 2 1/2 L. Now if we add demeclocycline (or even a loop diuretic) and the urine osmolality falls to 200 mOsm/L, then the allowable water intake would be roughly: 800 mOsm/200 mOsm/L + 1/2 L = 4 1/2 L. In general, an outpatient can manage to adhere to a fluid restriction of 2 L, but not much less. 4 1/2 liters should provide an adequate cushion of safety, but the patient should be followed with daily weights and measurement of his sodium concentration until the sodium concentration stabilizes.

18. Water restriction in patients with severe hyponatremia due to SIADH: A man with SIADH has a urine osmolality of 600 mOsm/L. (For purposes of this example, his urine osmolality is fixed at 600 mOsm/L. Actually, the urine osmolality may fluctuate during a 24-hour period.) His solute excretion is 600 mOsm/day. What approximate amount of fluid could he safely drink, and what could you do to improve this to a more reasonable amount?

Answer: Remember that this is an approximate way to get a rough idea of allowable fluid intake and will not give precise results. These patients need to be monitored with frequent measurements of the serum sodium concentration. The maximum amount of fluid the patient can drink per day is around 600/600 = 1L + insensible loss = 1 1/2 L per day! Any more water intake than this may result in progressive hyponatremia. The management of this patient is difficult because most patients are unable to restrict fluid intake to less than 2 L per day. This patient is at risk for developing dangerous hyponatremia.

What would happen if his renal solute excretion were increased to 900 mOsm/day by adding NaCl tablets? Answer: His allowable intake would be increased to 900/600 + 1/2 = 2 L. This is an improvement, but the patient is still at risk for serious hyponatremia. The other intervention is to decrease Uosm by the administration of demeclocycline or a loop diuretic. Decreasing Uosm to 300 from 600 would also increase the allowable water intake because the allowable water intake depends on the ratio of osmolal load excreted to the urine osmolality: 900/300 + 1/2 = 3 1/2 L.

Again, this is not a precise method to quantify allowable fluid intake, but a method to get a *rough* idea of allowable fluid intake in patients with chronic hyponatremia. Remember also that the *definitive* treatment of SIADH is correction, when feasible, of the underlying cause of the SIADH.

19. What would happen to a patient with hyponatremia caused by SIADH whose urine osmolality is fixed at 616 mOsm/L if 1 liter of 0.9% saline were given? Will the hyponatremia improve?

Answer: Patients with SIADH can regulate total body sodium and the size of the ECFV essentially like normal individuals (this is not entirely true but clinically is a good approximation). That is, they handle sodium normally. The 308 mOsm of NaCl contained in the liter of 0.9% saline is excreted. The volume of urine would be 308 mOsm/616 mOsm/L = .5 L. So what will happen to the other .5 L?

Answer: It is retained in the patient. Therefore, giving 1 liter of 0.9% saline to this patient will result in the net addition of .5 L of water to the patient. Sometimes, the diagnosis of the hyponatremia is unclear: It is not certain whether the patient has mild ECFV depletion or SIADH. The response of the serum sodium concentration to slow, *careful* administration of 0.9% saline with close monitoring of the sodium concentration and ECFV status may be helpful diagnostically. In ECFV depletion, the serum sodium concentration will usually begin to correct rapidly. In SIADH, the sodium concentration will usually not change much. Nevertheless, because of the possibility of ECFV overload, 0.9% saline administration is not recommended as a *routine* part of the diagnosis of hyponatremia.

CHAPTER 4. HYPERNATREMIA

Hypernatremia (serum sodium >145 mEq/L) generally results from a deficit of water. Most cases of hypernatremia require two things:

- Loss of water
- Failure to adequately replace the water loss

Hypernatremia will not generally develop in an alert person with an intact thirst mechanism and access to water, even though water losses are large. Inadequate intake of water is therefore a common denominator in nearly all cases of hypernatremia, regardless of the cause of water loss.

Because the sensitivity of thirst declines with aging, hypernatremia often occurs in elderly patients, especially in the setting of pulmonary or urinary tract infections. Hypernatremia also occurs in chronic debilitating illness and neurological disease. In evaluating patients with hypernatremia, one should carefully look for alterations in neurological status that are causing inadequate water intake.

The consequences of hypernatremia may be severe. Sodium concentration increases as water is lost. Water shifts out of cells to establish osmotic equilibrium, and brain cells shrink. The patient may become progressively lethargic, even comatose. Intracranial bleeding may arise, especially in children. The dehydrated, shrunken brain "hangs" by the meninges in the skull, which can tear the delicate bridging veins.

Causes of Hypernatremia

The causes of hypernatremia (see **Fig. 4-1**) may be divided into

- Extrarenal water loss
- Renal water loss
- Iatrogenic

The first two are due to water loss with inadequate replacement. Patients are generally dehydrated, and water replacement is indicated. On the other hand,

FIGURE 4-1. Causes of Hypernatremia

Extrarenal water loss (in addition to impaired thirst/intake)
  Insensible losses: fever, tachypnea, mechanical ventilation
  Sweat losses in hot environment
  GI losses: osmotic diarrhea (e.g. enteral tube feedings), acute infectious diarrhea
Renal water loss (in addition to impaired thirst/intake)
  Osmotic diuresis (Urine Osm >300 mOsm/L)
    Glucose
    Urea (e.g. enteral tube feedings)
    Mannitol
  Central diabetes insipidus (inadequate ADH)
    Head trauma
    Post-neurosurgical (craniopharyngioma, transsphenoidal surgery)
    Neoplastic (primary or metastatic)
    Sarcoidosis
    Histiocytosis X
    Meningitis/encephalitis
    Idiopathic
  Nephrogenic diabetes insipidus (inadequate renal response to ADH)
    Electrolyte disorders (hypercalcemia, hypokalemia)
    Drugs (lithium, demeclocycline)
    Recovery phase of acute renal failure
    Post urinary obstruction
    Chronic renal disease
Iatrogenic: Administration of hypertonic sodium

---

iatrogenic hypernatremia is due to administration of hypertonic saline or $NaHCO_3$, usually in the course of an acute, critical illness. Iatrogenic hypernatremia results from the *addition of hypertonic sodium,* rather than water loss.

Hypernatremia from Extrarenal Water Loss

The most common causes of hypernatremia due to extrarenal water loss include fever, profuse sweating, hyperventilation, including mechanical ventilation, and severe diarrhea. Patients with hypernatremia caused by extrarenal water loss often have decreased ECFVs as well, indicating deficits in total body sodium as well as water. The proportionally greater deficiency of water than of sodium leads to the increase in the serum sodium concentration.

Hypernatremia from Renal Water Loss

The hallmark of marked renal water loss is polyuria, defined as a urine volume greater than 3L/24 hours. The common defect in all cases of renal water loss is an inability of the kidney to conserve water appropriately. There are several important causes of renal water loss. The key to the evaluation of the patient with renal water loss is measurement of the urine osmolality.

Osmotic diuresis (urine osmolality >300 mOsm/L)

[handwritten: Appropriate dilution, something else is present though]

The excretion of an osmotic solute load by the kidney will obligate a certain loss of water. Polyuria is an important clue to the presence of an osmotic diuresis. How do we know if an osmotic diuresis is to blame for the hypernatremia? The osmotic load excreted by the kidney will be increased to more than 1200 mOsm/24 hours instead of the usual 600–900 mOsm/day. A helpful clue to the presence of an osmotic diuresis is the osmolality of a "spot" urine specimen: It will generally be in the range >300 mOsm/L and may have a sodium concentration in the range of 50–80 mEq/L. Patients with hypernatremia due to osmotic diuresis often manifest clinical signs of ECFV depletion. Typical clinical settings for the development of an osmotic diuresis are

- Poorly controlled diabetes mellitus (glucose is the osmotic agent)
- Hyperalimentation (central or enteral) with an increased load of urea from protein catabolism
- Mannitol administration for cerebral edema
- Administration of sodium-containing solutions with resultant sodium-induced diuresis.

Remember that acute administration of hypertonic mannitol or marked hyperglycemia may initially result in hyponatremia with hypertonicity. The patient is hypertonic, even though the serum sodium concentration is low because of transcellular shift of water. Osmotic diuresis resulting from the mannitol or hyperglycemia then results in progressive water loss and the sodium concentration increases. Ultimately, *hypernatremia* develops.

Diabetes insipidus (urine osmolality <150 mOsm/L)

[handwritten: Inappropriate dilution]

Diabetes insipidus results from the inability of the kidney to concentrate the urine appropriately due to either absence or deficiency of ADH (central diabetes insipidus) or unresponsiveness of the kidney to the effects of ADH (nephrogenic diabetes insipidus). The urine is inappropriately dilute and typically has a low sodium concentration in the presence of polyuria and a rising serum sodium concentration.

Central diabetes insipidus is due to deficiency of ADH (vasopressin). It is often associated with severe CNS structural lesions or infections, head trauma, or pituitary surgery (see **Fig. 4-1**). The onset of polyuria may be abrupt, with large urine volumes (5–10 L per 24 hours, for example). Patients sometimes describe a preference for ice water. Because the patients excrete large volume of dilute urine, a brief period of water restriction may result in a significant increase in the serum sodium concentration. The administration of vasopressin is used as a diagnostic test to establish the diagnosis of central diabetes insipidus. Vasopressin will result in a significant decrease in urine volume and an increase in urine osmolality.

Nephrogenic diabetes insipidus implies unresponsiveness of the kidney to ADH. It occurs in a number of interstitial renal diseases, electrolyte disorders, and in response to certain medications. The urine volumes are often less than those seen with central diabetes insipidus. Because the kidney is unresponsive to ADH, vasopressin administration will not result in a significant decrease in urine volume or a significant increase in urine osmolality.

Primary polydipsia (urine osmolality <150 mOsm/L)

Primary polydipsia is sometimes referred to as "psychogenic water drinking." Primary polydipsia is not a cause of hypernatremia but is mentioned here because these patients have polyuria and must be differentiated from patients who have polyuria due to diabetes insipidus. Patients with primary polydipsia do not develop hypernatremia; indeed they may develop hyponatremia if they have impaired ability to excrete excess water. These patients drink large volumes of water each day and often produce tremendous amounts of dilute urine. They do not develop hypernatremia when fluid restricted, however, and their urine osmolalities rise when water is restricted. It is important to remember that prolonged ingestion of vast amounts of water will sometimes produce a "washout" effect, removing osmoles from the medullary interstitium and thereby decreasing the medullary concentration gradient. The kidney's ability to produce a concentrated urine in response to water deprivation is therefore impaired. It may take several days for the medullary concentration gradient to be regenerated, and with it the ability to produce a maximally concentrated urine.

Iatrogenic Hypernatremia

Iatrogenic hypernatremia differs from the other categories of hypernatremia in that it is caused by *gain* of hypertonic sodium (usually $NaHCO_3$ or 3% saline) rather than *loss* of water. This usually occurs in the critical care unit or emergency room. Compromised renal function aggravates the problem. One 50 ml ampule of $NaHCO_3$ has 50 mEq of $NaHCO_3$—1000 mEq sodium per liter! One liter of 3% saline has 513 mEq/L sodium.

Diagnosis of Hypernatremia

Fig. 4-2 presents a diagnostic approach to the patient with hypernatremia. Although identification of the cause of water loss (or, in cases of iatrogenic hypernatremia, the source of hypertonic sodium) is important, attention must also be given to identification of the *reason why the patient is unable to replace the water losses* such as impaired thirst, altered mental status, or a primary neurological disorder.

FIGURE 4-2. Diagnosis of Hypernatremia

Step 1: Reason for water loss or sodium gain?
Increased insensible loss (fever, tachypnea)?
Sweat losses
Diarrhea
Renal water loss (>3L/24 hours of dilute urine)?
Administration of hypertonic sodium solutions (iatrogenic)?
Step 2: Reason for inadequate water intake regardless of source of water loss?
Impaired thirst
Altered mental status
Primary neurological disorder (stroke, infection, tumor)
No access to water
Step 3: Is polyuria present? Urine volume >3L/24 hours?
Urine Osmolality >300 mOsm/L (osmotic diuresis)
    Urea
    Glucose
    Mannitol
    Saline
Urine Osm <150 mOsm/L (diabetes insipidus)
    Response to vasopressin:
        No response: nephrogenic diabetes insipidus
        Urine Osm increases to >300 mOsm/L: central diabetes insipidus

Treatment of Hypernatremia

Severe ECFV depletion, especially when accompanied by hemodynamic compromise, is a first priority and should be corrected with 0.9% saline. Subsequent replacement fluids should be hypotonic. In general, the choice of which hypotonic fluid to administer is summarized in **Fig. 4-3.** Replacement of water deficits involves three steps:

- Calculation of the approximate water deficit
- Administration of the water replacement at a rate sufficient to correct the hypernatremia, but slowly enough to avoid cerebral edema, which is the major complication of overly rapid correction of hypernatremia
- Frequent rechecking of the sodium concentration to monitor therapy

The major complication of overly rapid correction of hypernatremia is cerebral edema. It is generally agreed that a safe rate of correction of hypernatremia is to decrease the serum sodium concentration initially by about 0.5–1 mEq/L per hour. Complete correction should not be accomplished for 36–72 hours. One approach would be to correct the serum sodium concentration to mildly hypernatremic levels, and then correct further at much slower rates.

The formula used to calculate the water deficit is:

$$H_2O \text{ deficit} = TBW \times ([Na^+_{(measured)}] - [Na^+_{(desired)}])/[Na^+_{(desired)}]$$

FIGURE 4-3. IV Fluids used in the treatment of hypernatremia

Solution	Indication	Cautions
D5W	Hypernatremia when it is desired to replace H_2O without Na^+. Each L delivers 1 L free H_2O. Efficient in correcting hypernatremia, unless glucosuria develops.	Will not be very effective at restoring ECFV in patients who are ECFV depleted. Hyperglycemia and glucosuria may result and may aggravate hypernatremia by causing osmotic diuresis. The patient should be monitored for hyperglycemia and glucosuria.
0.45 saline	Hypernatremia when it is desired to replace H_2O *and* Na^+. Each L delivers about 500 ml free H_2O. May be less efficient in correcting hypernatremia than D5W. Useful in hypernatremia with hyperglycemia.	Will be effective at restoring ECFV in patients who are also ECFV depleted. Hyperglycemia will not result. The solution of choice in severe hyperglycemia with hypernatremia.

where TBW is total body water, $[Na^+_{(measured)}]$ is the measured serum sodium concentration, and $[Na^+_{(desired)}]$ is the desired serum sodium concentration. This formula gives the approximate amount of water to be given in order to reduce the sodium concentration to the desired value. For example, if a 70 kg man has a sodium concentration of 170 and you desire to correct this to 160 mEq/L over the next 12 hours, then the amount of water to be given in this time will be:

$$H_2O \text{ deficit} = TBW \times ([Na^+_{(measured)}] - [Na^+_{(desired)}])/[Na^+_{(desired)}]$$
$$H_2O \text{ deficit} = 0.6 \times 70 \times (170-160)/160 = 2.6 \text{ liters}$$

This formula will give an approximate water deficit, but therapy must be reassessed by frequent determinations of the sodium concentration.

It is important to take into account ongoing losses of water into the total amount of water to be given over this period. For example, a patient might be expected to have insensible losses in the range 0.5–1 L per 24 hours, depending upon temperature and respiratory rate. Therefore, the total amount of water to be given to this 70 kg man over the next 12 hours, assuming an insensible loss of 0.5 L/24 hours, would be approximately 2.6 + 0.25 = 2.85 liters. Water may be administered as D5W, although the patient must be closely monitored for hyperglycemia and glucosuria. Many experts favor the enteral route for water administration whenever feasible because of potential problems related to hyperglycemia and the resulting osmotic diuresis.

Patients producing significant amounts of dilute urine due to diabetes insipidus or patients with diarrhea have significant ongoing water losses. These ongoing water losses must also be replaced. With fever, the insensible water loss increases by roughly 60–80 ml/24 hours for each degree Fahrenheit.

Exercises

1. A 79-year-old 60 kg man from a nursing home is admitted with fever, obtundation, and a urinalysis revealing pyuria and many bacteria. His temperature is 101.6 degrees, his blood pressure is 148/94, and his pulse 104. His mouth is dry, and he has poor skin turgor. His serum sodium is 184 mEq/L. The urine volume is .6 L/ 24 hours with urine osmolality 640 mOsm/L. What is the cause of the hypernatremia, and what should be done?

   Answer: The hypernatremia is most likely due to extrarenal losses accompanied by impaired thirst, causing inadequate water replacement. Infection is a common setting for hypernatremia in elderly debilitated patients. The total water deficit is:

   $$Total\ H_2O\ deficit = .5 \times 60 \times (184-140)/140 = 9.4\ L$$

   Because of the age of this patient, I have arbitrarily used the *rough* approximation $TBW = 0.5 \times$ body weight to avoid an overestimation of the total body water (.6 may be too high a fraction for body water in an elderly man). On the other hand, if $TBW = .6 \times$ body weight is used, the total water deficit is:

   $$Total\ H_2O\ deficit = .6 \times 60 \times (184-140)/140 = 11.3\ L$$

   The actual TBW may be somewhere in between .5 and .6 $\times$ body weight, and therefore, the total water deficit may be somewhere in between 9.4 and 11.3 L. We do not want to correct the sodium concentration too rapidly, but rather to correct it carefully to avoid cerebral edema. A good rate would be to correct the serum sodium concentration from 184 to 174 in the first 10 hours (1 mEq/L per hour correction of serum sodium).

   $$H_2O\ deficit = 0.5 \times 60 \times (184-174)/174 = 1.7\ L$$

   We need to keep an eye on all ongoing water losses. Ongoing losses need to be calculated into the final fluid volume of D5W to be given over the 10-hour period. Basal insensible loss over this 10-hour period will be perhaps 0.25 L. The amount of water replacement will be increased because of fever in this patient. The amount of D5W to be given to replace loss due to fever in addition to basal insensible loss would be (101.6–98.6) $\times$ 80 = 3 $\times$ 80 = 240 cc. Therefore, the amount of D5W to be given over the next 10 hours should be about 1.7 L + .25 L (basal insensible loss) + .24 L (insensible loss from fever) = 2.19 L. (I would round this to 2.2 L). The sodium concentration should be rechecked at 2 to 4-hour intervals to monitor therapy. Ignoring ongoing losses of water may lead to inadequate water replacement and prolongation of brain dehydration. Conversely,

71

overly rapid correction may lead to cerebral edema. The serum sodium concentration should therefore be checked at frequent intervals to monitor therapy. As for any patient receiving IV fluids, this patient should be weighed daily if possible, and have daily measurements of electrolytes, BUN and Cr.

2. You are called to evaluate a 34-year-old psychiatric patient because of polyuria. The patient is producing 6 liters per day of urine with an osmolality of 75 mOsm/L. What is the differential diagnosis?

   Answer: The differential diagnosis is between central or nephrogenic diabetes insipidus and primary polydipsia. This very low urine osmolality excludes osmotic diuresis. The serum sodium concentration is key: If it is elevated (>145 mEq/L), then central or nephrogenic diabetes insipidus is present, because primary polydipsia does not produce hypernatremia. The serum sodium concentration is often normal, however, regardless of the cause of the polyuria. This is because hypernatremia does not generally develop in someone who has an intact thirst mechanism *and* access to water, even in the presence of severe diabetes insipidus.

3. The serum sodium concentration in the patient in exercise 2 is 140 mEq/L. What do you do next?

   Answer: The next step is a water deprivation test, which consists of completely restricting water and measuring urine osmolality and serum sodium concentration serially. This test will distinguish between diabetes insipidus and primary polydipsia. The water deprivation test is done during the daytime under close observation because patients with severe diabetes insipidus can rapidly develop dangerous symptomatic hypernatremia when water is restricted. The patient with diabetes insipidus will continue to produce a dilute urine, and the serum sodium concentration will begin to rise.

   The patient with primary polydipsia will not develop hypernatremia because the osmolality of the urine will increase appropriately with water restriction, although the prolonged ingestion of large amounts of water will sometimes produce a "washout" effect of the medullary concentration gradient. In a patient with primary polydipsia, the kidney's ability to produce a concentrated urine in response to water deprivation may be initially impaired. It may take time for the medullary concentration gradient to be regenerated, and the ability to produce a concentrated urine may take several days to return.

   It is unnecessary to do a water restriction test in someone who already has hypernatremia because ADH should be maximally stimulated in a patient with hypernatremia. If a patient has hypernatremia, polyuria, and low urine osmolality, diabetes insipidus is present. *It is unnecessary and dangerous to restrict water in someone who is hypernatremic.*

4. The same patient as in exercises 2 and 3 continues to produce a dilute urine during water restriction. Serial determinations of urine osmolality are 75 mOsm/L, 68 mOsm/L, and 85 mOsm/L. The serum sodium concen-

tration rises to 146 mEq/L, and the patient complains of thirst. What is your diagnosis and what do you do next?

Answer: You have established the diagnosis of diabetes insipidus (central or nephrogenic). Vasopressin (ADH) is administered to distinguish between central and nephrogenic diabetes insipidus. In the patient with central diabetes insipidus, the urine osmolality will increase sharply following vasopressin administration, whereas the patient with nephrogenic diabetes insipidus will have little or no change in the urine osmolality.

5. Same patient. The urine osmolality does not change with vasopressin (ADH). What is the diagnosis?

Answer: The patient has nephrogenic diabetes insipidus. Drugs are among the causes shown in **Fig. 4-1.** Lithium is a common cause of nephrogenic diabetes insipidus among psychiatric patients.

6. Same patient. The urine osmolality increases to 620 mOsm/L. What is your diagnosis?

Answer: The patient has central diabetes insipidus. The causes are shown in **Fig. 4-1.**

7. A 24-year-old man presents with the complaint of frequent urination, excessive thirst, and a preference for ice water. Two weeks earlier, the patient had been admitted to the hospital for suspected meningitis. A 24-hour urine collection contains 4 L. The serum sodium is 148 mEq/L. A spot urine osmolality is 120 mOsm/L. What is your diagnosis?

Answer: If a patient has hypernatremia (serum sodium >145 mEq/L), polyuria, and low urine osmolality, then diabetes insipidus is present. In diabetes insipidus and primary polydipsia, the urine is dilute, below 150 mOsm/L. Our patient has hypernatremia and a low urine osmolality. Therefore, diabetes insipidus (either central or nephrogenic) is present. Primary polydipsia does not cause hypernatremia. The next step is to distinguish between central diabetes insipidus and nephrogenic diabetes insipidus. This is done by giving vasopressin and noting the response of the urine osmolality. Administration of vasopressin increases the urine osmolality in central diabetes insipidus, but not in nephrogenic diabetes insipidus.

In general, if a patient has hypernatremia, polyuria, and low urine osmolality, diabetes insipidus is present. It unnecessary and dangerous to restrict water in someone who is hypernatremic.

The approach to hypernatremia (**see Fig. 4-2**) does not include water deprivation. This is a very important point. The evaluation of a patient with polyuria and a *normal* serum sodium may require water restriction to rule out primary polydipsia. It is critical to distinguish between polyuria with hypernatremia (no water restriction should be done because it is dangerous) and polyuria with a normal serum sodium concentration (water restriction may be required to rule out primary polydipsia). **Fig. 4-2** summarizes the evaluation of hypernatremia, and therefore does not include reference to water restriction.

8. A 35-year-old patient comes to you with complaints of polyuria and poly-dipsia. A 24-hour urine collection contains 6 L. The serum sodium is 139 mEq/L. A spot urine osmolality is 120 mOsm/L. What is your approach? *Answer:* This patient has a low urine osmolality, suggesting that diabetes insipidus (either central or nephrogenic) or primary polydipsia is present. Because the serum sodium is not increased, a careful trial of water depri-vation is tried. This test should be done during the day so that the patient may be observed closely. In general, if the urine osmolality increases to the range 600 mOsm/L, then the diagnosis of primary polydipsia is made. If the urine osmolality remains low (below 200 mOsm/L) in response to dehydration, then diabetes insipidus is present. The patient with a urine osmolality between 200 and 600 mOsm/L will require further evaluation to determine the diagnosis. Vasopressin administration can distinguish between central diabetes insipidus and nephrogenic diabetes insipidus.

9. A 25-year-old 50 kg man with a history of post traumatic encephalopathy from a motor vehicle accident is admitted from a nursing home with fever, obtundation, and a urinalysis that reveals pyuria and 4+ bacteria. His serum sodium is 185 mEq/L. The urine volume is 0.7 L/ 24 hours with a urine osmolality of 710 mOsm/L. He is treated initially with 1 L per hour then 500 ml per hour of D5W. After 4 hours of therapy, he becomes more arousable, but after 12 hours, he has again become poorly responsive. His sodium concentration is 150 mEq/L. What has happened?
Answer: Cerebral edema secondary to overly rapid correction of the hyper-natremia. The brain adapts to a hypertonic ECFV by accumulating elec-trolytes, amino acids, and other osmoles that serve to increase the solute of the brain in order to "hold" water and prevent brain shrinkage as hyperna-tremia develops. The consequence of rapidly administering water is that water will rapidly enter brain cells, causing cerebral edema. It is generally agreed that a safe rate of correction of hypernatremia is about a 0.5–1 mEq/L per hour decrease in the serum sodium concentration initially, and that com-plete correction should not be achieved for at least 36–72 hours. One ap-proach would have been to correct the sodium from 185 to 175 mEq/L in 10 hours and then make further adjustments at a slower rate.

The formula used to calculate the water deficit is:

$$H_2O \text{ deficit} = TBW \times ([Na^+_{(measured)}] - [Na^+_{(desired)}])/[Na^+_{(desired)}]$$

This formula gives the amount of water to be given in order to reduce the sodium concentration to the desired value. For our patient:

$$H_2O \text{ deficit} = .6 \times 50 \times (185-175)/175 = 1.7 \text{ L}$$

This would be in addition to insensible losses of roughly .25 L (assuming no fever or hyperventilation) for a total of $1.7 + 0.25 = 1.95$ L over the next 10 hours.

10. A 79-year-old man (60 kg body weight) with a history of multi-infarct de-mentia is bedridden and requires enteral tube feedings. He is found to be

poorly arousable and has a respiratory rate of 26/minute. The following data are obtained: sodium 173 mEq/L, potassium 3.1 mEq/L, bicarbonate 18 mEq/L, chloride 137 mEq/L. The urine volume is <400 ml/24 hours, and a spot urine sodium concentration is 10 mEq/L. What is causing the hypernatremia?

Answer: Tube feedings can potentially cause hypernatremia by two means: (1) osmotic diuresis from urea produced by the metabolism of the amino acids provided by the tube feedings and (2) diarrhea. The urine output of < 400 ml/24 hours and the low urine sodium are against osmotic diuresis. Diarrhea leads to hypernatremia from loss of water in stool, hypokalemia from loss of potassium in stool, and metabolic acidosis from loss of bicarbonate in stool. The treatment is replacement of water, potassium, and (sometimes) bicarbonate. In this patient, the total free water deficit is approximately:

$$\textit{Total } H_2O \text{ deficit} = TBW \times ([Na^+_{(measured)}] - [Na^+_{(desired)}])/[Na^+_{(desired)}]$$

$$= .5 \times 60 \times (173{-}140)/140 = 7.1 \text{ L}$$

Because of the patient's age, .5 × body weight is used instead of .6 × body weight, although this is a rough approximation. The water deficit using .6 × body weight is 8.5 L. The actual water deficit may be in between .5 and .6 × body weight and therefore, between 7.1 and 8.5 L. It is important to get an idea of the general *range* of the water deficit in a patient such as this one: the calculations give only rough approximations of water deficit.

Suppose we wanted to correct him to a sodium concentration of 160 mEq/L in the first 14 hours, assuming 1 L/24 hours insensible loss because of the rapid respirations and therefore roughly 0.5 L in this 14-hour period:

$$H_2O \text{ to bring sodium to } 160 \text{ mEq/L}$$

$$= TBW \times ([Na^+_{(measured)}] - [Na^+_{(desired)}]) / [Na^+_{(desired)}]$$

$$= .5 \times 60 \times (173{-}160)/160$$

$$= 2.4 \text{ L}$$

Now add roughly 0.5 L for insensible loss because the patient is tachypneic. The volume of water, which could be given as D5W over the next 14 hours, would be about 2.4 + 0.5 = 2.9 L. One of the keys to successful correction of the patient's hypernatremia is close observation for ongoing water loss in the stool. If he continues to have diarrhea, the patient will most likely need to have increased water replacement to account for the stool losses. Therapy must be monitored closely with frequent serum sodium determinations. Potassium replacement is discussed in Chapter 5 and metabolic acidosis in Chapter 7.

11. You are called to evaluate a comatose 44-year-old man in the intensive care unit because of polyuria. The patient is producing 6 L per day of

urine. The serum sodium concentration is 147 mEq/L. A spot urine osmolality is 320 mOsm/L. The urine osmolality of a 6 L 24-hour collection is 315 mOsm/L. What do you do?

Answer: The urine osmolality is immediately helpful. It is > 300 mOsm/L, and therefore we suspect an osmotic diuresis. The total 24-hour osmolal excretion is 1890 mOsm, which is well above 1200 mOsm. Both the spot urine osmolality of 320 mOsm/L and the 24 hour osmolal excretion of 315 mOsm/L $\times$ 6 L = 1890 mOsm support osmotic diuresis as the cause of the polyuria and the hypernatremia. In the ICU setting, possible osmotic agents are:

• Urea, generated by protein catabolism and hyperalimentation
• Glucose, in uncontrolled diabetes
• Mannitol
• NaCl and other electrolytes from IV solutions

An important source of solute is the administration of saline solutions. Many times, a patient is appropriately mobilizing excess water and electrolytes that were administered during the course of an acute illness (septic shock, for example). The patient experiences a saline diuresis during the recovery phase of an acute illness, which is interpreted as an abnormal diuresis, although the patient will generally not develop severe hypernatremia. Polyuria is sometimes also observed in the recovery phase of acute renal failure and following relief of urinary obstruction (postobstructive diuresis).

CHAPTER 5. HYPOKALEMIA

The clinical consequences of significant hypokalemia include:

- Neuromuscular manifestations (weakness, fatigue, paralysis, respiratory muscle dysfunction, rhabdomyolysis)
- Gastrointestinal manifestations (constipation, ileus)
- Nephrogenic diabetes insipidus
- ECG changes (prominent U waves, T wave flattening, ST segment changes)
- Cardiac arrhythmias (especially with concurrent digitalis)

Causes of Hypokalemia

The causes of hypokalemia (see **Fig. 5-1**) are divided into:

- Spurious hypokalemia
- Redistribution hypokalemia
- Extrarenal potassium loss
- Renal potassium loss

Spurious Hypokalemia

In spurious hypokalemia, the potassium concentration is not really low.

- Marked leukocytosis (>100,000) rarely may produce spurious hypokalemia if the blood tube is allowed to sit at room temperature. White cells may simply take up the potassium in the blood specimen.
- A dose of insulin right before blood drawing could cause temporary movement of potassium into cells in the blood tube and falsely lower the serum potassium. The magnitude of the fall in potassium is generally small (around 0.3 mEq/L).

Redistribution Hypokalemia

Redistribution hypokalemia is caused by the entry of potassium into cells. Only a small amount of total body potassium is located in the extracellular compartment. Consequently, a small shift of potassium from the extracellular

FIGURE 5-1. Causes of Hypokalemia

Spurious
  Marked leukocytosis (WBC >100,000)
Redistribution (potassium shifts into cells)
  Insulin administration just prior to blood drawing
  Alkalemia
  Beta$_2$ adrenergic activity/agents
  Theophylline toxicity
  Familial hypokalemic periodic paralysis
  Hypokalemic periodic paralysis with thyrotoxicosis
  Factor replacement in megaloblastic anemia
Hypokalemia caused by extrarenal loss (urine potassium <20 mEq/24 hours)
  Diarrhea
  Laxative abuse
  Villous adenoma of recto-sigmoid colon
  Sweat losses
  Fasting/inadequate intake
Hypokalemia caused by renal loss (urine potassium >20 mEq/24 hours)
  With metabolic acidosis
    Type I (distal) renal tubular acidosis
    Type II (proximal) renal tubular acidosis
    Diabetic ketoacidosis
    Carbonic acid anhydrase inhibitors
    Ureterosigmoidostomy
  With metabolic alkalosis
    Vomiting/nasogastric suction
    Diuretic therapy
    Post hypercapnea
    Mineralocorticoid excess syndromes
    Gitelman's syndrome
    Barrter's syndrome
  With no specific acid-base disorder
    Recovery phase of acute renal failure
    Post obstructive diuresis
    Osmotic diuresis
    Saline administration
    Magnesium depletion
    Aminoglycoside antibiotics
    Cis platinum
    Sodium penicillins
    Some leukemias

space to the intracellular space can cause a large change in plasma potassium concentration.

Alkalosis. Potassium concentration decreases because potassium shifts into cells. One very rough guide to the magnitude of the shift is that the serum potassium concentration falls by about 0.3 mEq/L for each 0.1 increase in pH. However, alkalosis often results from disorders that deplete total body potassium. Therefore, true depletion of total body stores is usually present as well as redistribution hypokalemia when metabolic alkalosis is present.

Increased beta$_2$-adrenergic activity. Stimulation of beta$_2$-adrenergic receptors shifts potassium into cells by increasing the activity of sodium-potassium ATPase. In states of sympathetic hyperresponsiveness, as occur in myocardial infarction, delerium tremens, major head trauma, and administration of beta-sympathomimetic agents during the treatment of asthma, there may be a transient shift of potassium into cells and a decrease in serum potassium concentration.

Theophylline toxicity may shift potassium into cells. The mechanism is unknown, but hypokalemia may aggravate the serious arrhythmias that sometimes occur in severe theophylline toxicity.

Familial hypokalemic periodic paralysis (autosomal dominant) is an uncommon cause of hypokalemia, due to a transcellular shift of potassium. This disorder is associated with recurrent episodes of flaccid paralysis that begin in childhood and are accompanied by hypokalemia: Serum potassium is often less than 3.0 mEq/L during periods of paralysis.

Hypokalemic periodic paralysis with thyrotoxicosis is associated with paralytic episodes clinically similar to those of the familial form. This disorder is seen as a complication of thyrotoxicosis, especially in patients of Asian heritage.

Factor replacement therapy for severe megaloblastic anemias results in the rapid assimilation of potassium into red blood cells as they are produced. This can cause a significant fall in serum potassium concentration and require potassium replacement. This fall typically occurs approximately 2 days after beginning therapy for the anemia.

Extrarenal Potassium Depletion

Total body potassium depletion may result from either renal or extrarenal potassium loss. In most cases of extrarenal potassium loss, renal potassium conservation is present (urine potassium <20 mEq/24 hours in a patient with hypokalemia).

Severe secretory **diarrhea** results in loss of potassium with HCO_3^- resulting in hypokalemia and metabolic acidosis. **Villous adenoma** of the rectosigmoid colon may present with hypokalemia, often associated with metabolic acidosis. **Chronic laxative abuse** can result in severe potassium depletion and metabolic alkalosis. **Sweat** has a potassium concentration of roughly 9 mEq/L; athletic training or hard work in the hot sun may produce up to 12 L/day of sweat and significant potassium deficits. **Fasting or inadequate intake** alone will result in only moderate potassium depletion because of the ability of the kidney to conserve potassium. The entire deficit is generally not more than 300 mEq if poor intake alone is involved.

Renal Potassium Depletion

Many of the disorders causing renal potassium loss (urine potassium >20 mEq/24 hours in a patient with hypokalemia) are also associated with acid-base disorders. Therefore, it is customary to classify the numerous causes of renal potassium loss according to whether they typically occur together with

• Metabolic acidosis
• Metabolic alkalosis
• No specific acid-base disorder

Renal hypokalemia with metabolic acidosis

The causes of renal potassium depletion with metabolic acidosis are:

• Renal tubular acidosis type I (distal) and type II (proximal)
• Diabetic ketoacidosis
• Carbonic anhydrase inhibitor therapy
• Ureterosigmoidostomy.

These are discussed more completely in Chapter 7 on metabolic acidosis.

Renal potassium depletion with metabolic alkalosis

Metabolic alkalosis is almost always associated with hypokalemia because virtually all of the conditions that cause metabolic alkalosis also lead to potassium depletion (see **Fig. 5-1**). In many types of metabolic alkalosis, the excess HCO_3^- acts as a poorly reabsorbable anion and "carries" more sodium to the collecting tubule, leading to increased sodium-potassium exchange and urinary potassium loss. This can be especially important in states of ECFV depletion that are associated with high levels of aldosterone. The causes and pathophysiology of metabolic alkalosis are discussed in Chapter 8.

Renal hypokalemia with no specific acid-base disorder

Recovery from acute renal failure, postobstructive diuresis, and osmotic diuresis can all lead to renal potassium loss and significant potassium depletion. The most important mechanism in these conditions is increased delivery of sodium to the collecting tubule resulting in increased potassium secretion.

A very important and often overlooked cause of renal potassium loss is **magnesium depletion.** Magnesium depletion induces renal potassium loss by complex mechanisms. It is very difficult to correct the potassium deficiency until the coexisting magnesium deficit is corrected: The urinary potassium loss will continue despite large replacement doses of potassium. Magnesium depletion should always be suspected when there is hypokalemia with persistent renal potassium loss.

Penicillins can cause renal potassium loss by acting as poorly reabsorbable anions, which thereby increase distal sodium delivery and sodium-potassium exchange. **Gentamicin** and **cisplatin** have direct tubular toxic effects that induce potassium loss. Certain **leukemias** are occasionally associated with renal potassium loss.

Diagnosis of Hypokalemia

Total body potassium depletion may result from either renal or extrarenal potassium loss. Very often the source of the potassium loss (renal versus extrarenal) is evident upon careful history. If the source of potassium loss is unclear, the most useful test to distinguish between renal and extrarenal loss is the measurement of 24-hour urinary potassium excretion. A 24-hour determination of >20 mEq/24 hours in the presence of hypokalemia implies that renal potassium loss is the cause of the hypokalemia. Spot urine potassium determinations are less useful because hypokalemia induces a water diuresis in many patients; the excess water dilutes the specimen and misleadingly lowers the urine potassium concentration.

Alternatively, some recommend the spot urine potassium/creatinine ratio for determination of renal versus nonrenal source of hypokalemia. A value of >20 mEq/gram supports the diagnosis of renal loss of potassium.

Treatment of Hypokalemia

Oral potassium replacement is preferred to intravenous. Although several salts of potassium are available, potassium chloride is used most frequently. Potassium chloride is used to correct the hypokalemia in cases of metabolic alkalosis with ECFV depletion and most other causes of hypokalemia. In type I and type II renal tubular acidosis and in diarrhea, potassium bicarbonate or potassium citrate (citrate is converted to bicarbonate in the liver) may be used to replace both potassium and bicarbonate.

Intravenous administration of potassium is appropriate in patients with profound, life-threatening hypokalemia and in patients who are unable to tolerate potassium by mouth. Intravenous administration is potentially dangerous because of the risk of severe, acute hyperkalemia. Potassium is irritating to veins, and concentrations more than 30 mEq/L and rates of administration more than 10 mEq/hr are not generally recommended. In emergency situations (for example, profound hypokalemia with metabolic acidosis) potassium may be given at higher rates and in higher concentrations via catheters in large veins in a closely monitored setting, with frequent determinations of potassium concentration.

Estimation of Total Body Potassium and Potassium Deficits

With regard to potassium replacement, it is especially important to remember that not everyone is a 70 kg man! Given the tiny proportion of total body potassium in the extracellular space, the serum potassium concentration should be used only as a *rough* guide to estimate the magnitude of potassium deficiency. The total amount of potassium normally contained in the body is proportional to muscle mass and body weight. Muscle mass declines with age, and is generally greater in men than in women. Depending upon age, gender, and body weight, there is a striking variation among individuals in total body potassium content. Consequently, it is very important to consider these patient characteristics when assessing a patient's total body potassium, in order to prevent serious errors in potassium administration. In general, each 1 mEq/L decrease in potassium concentration reflects a deficit of 150–400 mEq in total body potassium. The 150 mEq deficit for a 1 mEq/L decrease in potassium concentration might apply to an elderly woman with small muscle mass, whereas the 400 mEq deficit for a 1 mEq/L decrease in potassium concentration might apply to a 20-year-old man with a large muscle mass.

In a stable patient with a moderate degree of hypokalemia (potassium concentration $>$ 3.0 mEq/L) and no ongoing losses of potassium, potassium should be given gradually, in divided doses of oral replacement over a period of days, with frequent determinations of potassium concentration.

Exercises

1. Review question. How much potassium is there in the ECFV of a 70 kg man?
   Answer: The very delicate nature of the transcellular distribution of potassium is illustrated by the following calculation:
   TBW = .6 $\times$ 70 kg = 42 L
   ECFV = 1/3 $\times$ 42 L = 14 L
   Potassium concentration in ECFV: 4.0 mEq/L
   Total potassium in ECFV: 4.0 mEq/L $\times$ 14 L = 56 mEq

2. Review question. How much potassium is there in the ECFV of a 40 kg woman?
   Answer:
   TBW = .5 $\times$ 40 kg = 20 L
   ECFV = 1/3 $\times$ 20 L = 6.7 L
   Potassium concentration in ECFV: 4.0 mEq/L
   Total potassium in ECFV: 4.0 mEq/L $\times$ 6.7 L = 26.8 mEq
   The calculated amount of potassium in the entire ECFV is about *one* supplemental 20 mEq dose of KCl!

3. A 60-year-old woman with chronic lymphocytic leukemia has a serum potassium of 3.0 mEq/L. The lab reports to you that the blood specimen remained on the intake counter for three hours. Her white blood cell count is 150,000. What do you do?
   Answer: Repeat the test. Spurious hypokalemia may be present.

4. What would be the approximate serum potassium concentration of a patient with a severe alkalosis, pH 7.7, and a serum potassium 2.0 mEq/L if the pH were corrected to normal without giving any potassium replacement?

Answer: One very rough guide to the magnitude of the shift in potassium due to metabolic alkalosis is that the potassium falls by about 0.3 mEq/L for each 0.1 increase in pH. The patient would have a corrected serum potassium of around $2.0 + 3 \times 0.3 = 2.9$ mEq/L. Remember that this is only an approximation. The point is that in states of metabolic alkalosis the potassium deficit may not be as severe as might initially be suspected based on the potassium concentration. Another point is also important: Some types of metabolic acidosis have the opposite effect to that of alkalemia of shifting potassium out of cells. A patient with *metabolic acidosis* and hypokalemia may have a more *severe* deficit than might initially be suspected based on the potassium concentration. A patient with metabolic acidosis and severe hypokalemia is a medical emergency.

5. A man who is homeless has been eating poorly for the past 20 days. A colleague makes the remark that "his nutritional status can't be all that bad because his potassium concentration is 3.8 mEq/L." What is your answer?
   Answer: The potassium of 3.8 mEq/L is not a good indicator of this man's poor nutritional status. In about 3–5 days the kidney adjusts to a sudden decrease in potassium intake and begins effective renal potassium conservation. The normal kidney conserves potassium well in states of low intake by decreasing the urine potassium to less than 20 mEq/24 hours. After 24 hours, significant net losses should not continue, and the patient will not be in significant negative potassium balance unless potassium intake is significantly below 20 mEq/day. That is, with an intake of only 20 mEq/day, the kidney should be able to conserve losses to below 20 mEq/24 hours, keeping net potassium loss close to zero.

6. A 52-year-old man with a long history of alcoholism presents with a 4-day history of nausea and vomiting, midepigastric abdominal pain, and muscle weakness. The physical exam is compatible with marked volume depletion. Laboratory data: sodium 130 mEq/L, potassium 2.3 mEq/L, chloride 74 mEq/L, HCO_3^- 40 mEq/L, calcium 7.2 mg/dl, amylase 1125. An arterial pH is 7.52. What do you think about the origin of his hypokalemia?
   Answer: The metabolic alkalosis (high pH and high $[HCO_3^-]$) is secondary to the protracted vomiting. The hypokalemia associated with vomiting results from *urinary* potassium losses, because
   • ECFV depletion stimulates aldosterone
   • The high HCO_3^- concentration acts as a poorly reabsorbable anion, "carrying" sodium to the collecting tubule
   The combination of high aldosterone levels and increased collecting tubule sodium delivery results in increased potassium excretion. Notice that the calcium is low. The combination of hypokalemia and hypocalcemia in an alcoholic patient with protracted vomiting should bring to mind magnesium depletion as a complicating electrolyte abnormality, in addition to pancreatitis. Rhabdomyolysis also occurs in this patient population and can cause hypocalcemia.

7. The same patient as in exercise 6 after 3 days of IV potassium replacement remains with a potassium concentration of 2.6 mEq/L. What has happened?

Answer: The patient most likely has magnesium depletion leading to renal potassium wasting. The inability of the kidney to conserve potassium in this setting makes it almost impossible to correct the potassium deficit until the magnesium deficit is replaced. Magnesium replacement would also be expected to help correct the hypocalcemia of such a patient.

CHAPTER 6. HYPERKALEMIA

Severe hyperkalemia may be a medical emergency requiring immediate treatment, depending upon the nature of any ECG abnormalities. Clinical manifestations of hyperkalemia usually occur when the potassium concentration is >6.5 mEq/L and include:

- Neuromuscular signs (weakness, ascending paralysis, and respiratory failure)
- Typical progressive ECG changes with increasing potassium concentration: peaked T waves, flattened P waves, prolonged PR interval, idioventricular rhythm and widened QRS complex with deep S waves. Finally, a "sine wave" pattern develops, followed by ventricular fibrillation.

The cardiac changes may occur suddenly and without warning.

Causes of Hyperkalemia

Pseudohyperkalemia

In **pseudohyperkalemia,** the potassium concentration is artifactually high (see **Fig. 6-1**). Spurious causes of hyperkalemia, in addition to simple lab error, consist of marked thrombocytosis (platelet count >1,000,000); severe leukocytosis (white blood cell count >200,000); mononucleosis; ischemic blood drawing/ hemolysis during blood drawing; and a rare condition known as familial pseudohyperkalemia, in which potassium "leaks" out of red blood cells while the blood is waiting to be analyzed.

Redistribution Hyperkalemia

Redistribution hyperkalemia is caused by potassium transiently leaving cells, thereby raising the serum potassium concentration. Total body potassium need not be increased for redistribution hyperkalemia to develop. Only a small amount of potassium is located in the extracellular compartment (about 56 mEq in a 70 kg man, compared to a total body potassium content of around 4200 mEq/L for this individual). Consequently, a relatively small shift of potassium from the intracellular space to the extracellular space can cause a large increase in plasma potassium concentration.

FIGURE 6-1. Causes of Hyperkalemia

Pseudohyperkalemia
  Hemolysis during blood drawing
  Excessive fist clenching with tourniquet during blood drawing
  Platelets >1,000,000
  WBC >200,000
  Mononucleosis
  Familial pseudohyperkalemia (potassium efflux from cells)
Redistribution (potassium shifts out of cells)
  Acidosis (metabolic and respiratory)
  Hypertonic states
  Massive digitalis overdose
  Autosomal dominant hyperkalemic periodic paralysis
Aldosterone deficiency/ unresponsiveness
  Primary adrenal failure (autoimmune, TB, hemorrhage, tumor infiltration)
  Syndrome of hyporeninemic hypoaldosteronism (SHH)
    Accounts for many cases of unexplained hyperkalemia
    GFR is generally >20%
    May have an associated non anion gap metabolic acidosis
    Caused by a variety of interstitial renal diseases
    Diabetes is the most common cause
  Tubular unresponsiveness to aldosterone
    Caused by a variety of interstitial renal diseases
    Very similar to SHH but does not respond to fludrocortisone
Renal failure
  GFR is typically reduced to <10–20% of normal.
  Hyperkalemia may develop rapidly from exogenous potassium in patients with renal
    failure.
Drugs that can cause or aggravate hyperkalemia (multiple mechanisms)
  Potassium sources
    IV potassium solutions
    Potassium penicillin
    Potassium citrate
    Salt substitutes
  Redistribution hyperkalemia
    Arginine and lysine HCl (rarely used)
    Beta$_2$-adrenergic blockers
    Succinylcholine
    Digitalis (massive overdose)
    Hypertonic mannitol (osmotic effect)
  Aldosterone deficiency/unresponsiveness
    Nonsteroidal anti-inflammatory drugs (NSAIDs)
    Angiotensin-converting enzyme inhibitors
    Angiotensin II receptor antagonists
    Heparin
    Cyclosporine
    FK 506
    Aminoglutethimide
    Potassium-sparing diuretics
    Trimethoprim
    Pentamidine
    Nefamostat mesilate

Both **metabolic acidosis** and **respiratory acidosis** can shift potassium out of cells in exchange for hydrogen ion. There are some very rough approximations to estimate the magnitude of potassium shift caused by metabolic acidosis. For many types of metabolic acidosis the potassium will increase by roughly 0.7 mEq/L for every 0.1 decrease in pH. For respiratory acidosis the potassium will increase by roughly 0.3 mEq/L for every 0.1 decrease in pH. Acidoses caused by organic acids such as lactic acidosis and ketoacidosis do *not* generally lead to significant redistribution hyperkalemia for reasons that are complex.

Beta-adrenergic blocking agents can lead to modest (0.1–0.2 mEq/L) increases in potassium concentration, secondary to redistribution.

Hypertonic states may be associated with hyperkalemia. Once the hypertonicity is corrected, the potassium concentration may fall rapidly. A normal serum potassium in the presence of a hyperosmolal state implies potassium depletion.

Massive overdose with digitalis results in redistribution hyperkalemia secondary to inhibition of cell membrane sodium-potassium ATPase. The neuromuscular blocking agent **succinylcholine** increases the potassium permeability of muscle cells and can cause a mild increase in potassium of about 0.5 mEq/L in normal individuals. More severe hyperkalemia may be induced in burn patients and in patients with chronic neuromuscular disease.

Familial hyperkalemic periodic paralysis is an uncommon cause of hyperkalemia. This autosomal dominant disorder is associated with recurrent episodes of flaccid paralysis with hyperkalemia. Serum potassium is often in the range 6.0–8.0 mEq/L. The attacks may be precipitated by the intake of a high potassium diet or exposure to cold, and may last from minutes to hours.

Hyperkalemia Secondary to Impaired Potassium Excretion

The majority of cases of hyperkalemia secondary to true excess of total body potassium are due to a defect in renal potassium excretion in the presence of ongoing potassium intake. The impaired renal potassium excretion is due to one or both of the following:

• Aldosterone deficiency or tubular unresponsiveness to aldosterone
• Renal failure (reduced GFR)

Expressed simply: Hyperkalemia with excess total body potassium is generally due to either a problem with aldosterone or a problem with GFR.

Aldosterone Deficiency and Aldosterone Unresponsiveness (Type IV RTA)

True potassium excess (increased total body potassium) due to renal retention of potassium will develop if there is a deficiency of aldosterone, or tubular unresponsiveness to the kaliuretic effects of aldosterone. The syndrome of hyperkalemia secondary to a variety of disorders which cause aldosterone deficiency or unresponsiveness to aldosterone is called **type IV renal tubular acidosis** (see **Figs. 6-1** and **7-2**).

Primary adrenal failure secondary to autoimmunity, tuberculosis, tumor infiltration, or hemorrhage will lead to both cortisol and aldosterone deficiency. On the other hand, pituitary failure will affect cortisol but *not* aldosterone, so that pituitary failure is not an important cause of hyperkalemia.

The **syndrome of hyporeninemic hypoaldosteronism (SHH)** is a common cause of hyperkalemia. It is characterized by low plasma renin activity, low plasma aldosterone, and hyperkalemia. There may be an associated mild metabolic acidosis. Mild to moderate renal insufficiency is typical, but the GFR is often >20% of normal. This syndrome is seen in a variety of renal disorders, but the most common cause is diabetes mellitus. The treatment is loop diuretics or loop diuretics plus mineralocorticoid replacement.

Tubular unresponsiveness to aldosterone occurs with a number of chronic renal disorders. The syndrome is quite similar to hypoaldosteronism in clinical presentation, but plasma renin activity and plasma aldosterone are not low. There may be mild renal insufficiency present, but the GFR is usually not low enough to explain the hyperkalemia on the basis of renal failure alone (the GFR is usually >20% of normal). These patients do not respond to mineralocorticoid replacement.

A number of **drugs** can cause hyperkalemia by interfering with the production of aldosterone or by blocking the kaliuretic effects of aldosterone. Commonly used drugs which can cause hyperkalemia by these mechanisms include amiloride, spironolactone, triamterene, trimethoprim, heparin, nonsteroidal antiinflammatory drugs, and angiotensin-converting enzyme inhibitors.

Renal Failure

The normal kidney adjusts its excretion of potassium to a wide range of potassium intake, thereby maintaining a constant total body potassium and ECF potassium concentration. Excretion is as low as 10 mEq/day during states of extreme potassium conservation to as high as 10 mEq *per kg body weight/* 24 hours. The upper limit of potassium excretion is *roughly* proportional to the GFR. If the GFR is 100% of normal, the maximum amount of potassium which can be excreted in one day is roughly 10 mEq per kg body weight. This is about $70 \times 10 = 700$ mEq in a 70 kg person. If the GFR is reduced to 50% of normal the maximum amount of potassium that can be excreted in one day falls to approximately $50\% \times 700 = 350$ mEq. This is a rough approximation of maximum potassium excretion because compensatory renal potassium secretory mechanisms will increase potassium excretion, and stool potassium losses also increase as the body defends itself against hyperkalemia. If the GFR is further reduced to 20% of normal, the maximal potassium excretion would fall to the range of about 140 mEq/day (20% of 700 mEq/day).

The average diet has about 1 mEq of potassium per kg body weight, which amounts to about 70 mEq/day in a 70 kg person. For a diet containing 70 mEq/day, the GFR would need to be reduced to approximately $70/700 = 10\%$ of normal before hyperkalemia develops. In fact, the GFR is usually below this level when hyperkalemia develops based upon usual dietary intake. Hyperkalemia may develop at less profound levels of renal failure if the potassium intake is increased or if there is a hidden potassium load. For example, a person with a diet high in potassium would develop hyperkalemia with less impairment of the GFR. Using a rough estimate of maximum potassium excretion, a patient with a GFR 15% of normal would develop hyperkalemia if dietary potassium is over the range of $15\% \times 700 = 105$ mEq/day. As mentioned above: this is only a rough approximation of maximum potassium excretion because compensatory renal potassium secretory mechanisms will increase potassium excretion, and stool potassium losses also increase as the body defends itself against hyperkalemia.

The clinical point is that if a patient has mild to moderate renal failure and hyperkalemia, the hyperkalemia should not be simply ascribed to renal failure alone. A vigorous search for other causes of hyperkalemia is needed.

Drugs

A number of drugs can cause hyperkalemia or aggravate existing hyperkalemia by a variety of mechanisms (see **Fig. 6-1**). These drugs should be used with caution, if at all, in patients who are predisposed to hyperkalemia.

A Few Comments about the Patient "At Risk" for Hyperkalemia

Impaired potassium excretion places a patient at risk for acute hyperkalemia should excessive potassium be supplied or a medication capable of causing hyperkalemia be prescribed. Therefore, a patient with a normal potassium concentration who has aldosterone deficiency, tubular unresponsiveness to aldosterone, or renal failure is at risk for developing hyperkalemia. Medications capable of producing hyperkalemia must be avoided in such a patient.

It is also important to note that the causes of hyperkalemia may be additive. That is, a given patient may have more than one cause of hyperkalemia acting to elevate the serum potassium concentration. Therefore, all potential causes of hyperkalemia should be systematically evaluated in every hyperkalemic patient.

Diagnosis and Treatment of Hyperkalemia

When hyperkalemia is present, diagnosis and treatment must begin simultaneously (see **Figs. 6-2** and **6-3**). There may be no time to carefully ponder the diagnosis of hyperkalemia in a patient with severe hyperkalemia. Potassium administration is stopped at once. This may sound obvious, but overlooking this simple first step could be disastrous. An ECG is obtained immediately and inspected for evidence of hyperkalemia. The urgency of therapy depends upon the presence or absence of important ECG changes: Severe hyperkalemia causing significant ECG changes is a medical emergency. Peaked T waves are the earliest ECG manifestation of hyperkalemia and confirm the

Remember that there is often more than one cause of a patient's hyperkalemia. Impaired renal potassium excretion is usually present.

Step 1: Stop all administration of potassium (oral, enteral or IV).

Step 2: Obtain a stat ECG.

Peaked T waves confirm that true hyperkalemia is present.

More severe ECG manifestations of hyperkalemia indicate emergency therapy with intravenous infusion of calcium (carefully in a patient receiving digitalis) to counteract the cardiac effects of hyperkalemia.

Step 3: Quickly seek possible "hidden" sources of potassium.

Potassium penicillin

Salt substitutes (many contain KCl)

Hemolysis

Gastrointestinal hemorrhage

Rhabdomyolysis

Burns

Major surgery

Drugs that cause or aggravate hyperkalemia

Step 4: Send stat repeat potassium (drawn without tourniquet to reduce risk of hemolysis).

Step 5: Find the underlying cause of hyperkalemia (see **Fig. 6-1**).

Is pseudohyperkalemia present?

Thrombocytosis

Leukocytosis

Is the sample hemolyzed?

Is redistribution hyperkalemia present?

Is there aldosterone deficiency/unresponsiveness?

Many times, these states are associated with a mild chronic hyperkalemia, which suddenly worsens in response to a potassium load. A careful look at previous records may disclose evidence for diabetic renal disease or chronic interstitial renal disease.

Is renal failure present?

GFR <20% of normal

Hyperkalemia may occur with less severe renal failure in the presence of large potassium loads.

---

elevated potassium concentration. More serious ECG manifestations are flattening of the P waves, prolongation of the PR interval, and widening of the QRS complex with the development of deep S waves. Finally, a "sine wave" pattern develops, followed by ventricular fibrillation and cardiac arrest.

When a significantly elevated potassium is discovered, potassium should be remeasured, drawing blood without using a tourniquet if possible, to confirm the result. Determination of the plasma potassium as opposed to the serum potassium (which is the usual test done by the clinical lab) may be helpful in excluding spurious hyperkalemia from a markedly increased number of platelets or leukocytes, because potassium is released when the blood clots in the serum sample.

The increase in total body potassium needed to produce significant hyperkalemia is not large. For example, increasing the potassium from 5.0 to

FIGURE 6-3. Agents Used in the Treatment of Hyperkalemia

Agent	Mechanism	Cautions
Calcium gluconate 10 ml of 10% solution (1 gram) IV slowly over 5–10 minutes.	Temporarily (1 hour) antagonizes cardiac effects of hyperkalemia while more definitive therapy is begun.	May induce digitalis toxicity: give only with great caution to patients receiving digitalis. May precipitate if given with solutions containing $NaHCO_3$.
Glucose 100 ml of 25% solution (25 gm) with 10 U regular insulin.	Temporarily translocates potassium into cells. Effect occurs within 30–60 minutes and lasts about 1 hour.	May induce hyperglycemia. If patient is already hyperglycemic, glucose infusion is not required.
Beta 2 agonists	Temporarily translocate potassium into cells.	Potentially dangerous in the setting of coronary artery disease.
1 standard ampule $NaHCO_3$ (50 mEq) IV over 5–10 minutes in patient who is acidemic.	May be indicated when acidosis is present with hyperkalemia (see Chapter 7).	May cause ECFV overload, alkalosis, hypernatremia. May produce seizures/tetany if hypocalcemia is present. Will precipitate with calcium solutions.
Sodium polystyrene sulfonate (Kayexalate®) 15–30 grams in sorbitol orally.	Each gram may remove about 1 mEq potassium from the body in exchange for 1–2 mEq of sodium.	ECFV overload. Intestinal necrosis from sorbitol reported in surgical postop patients.
Sodium polystyrene sulfonate (Kayexalate®) powder 50 grams in 200 ml 20% D/W as retention enema.	Removes potassium from the body in exchange for sodium.	Sodium retention and ECFV overload. Oral route is preferred. Cleansing enemas required to remove the resin. Occasional cases of colonic necrosis.
Hemodialysis	Removes potassium from the body.	Should be used in patients with renal failure when more conservative treatments have been tried without success.

6.0 mEq/L may only require 100–200 mEq of potassium. Therapy is directed in three ways:

- Emergency treatment to counteract the cardiac effects of hyperkalemia when advanced ECG changes are present. This is done by the IV infusion of calcium.

- Temporizing treatment to acutely drive potassium into cells: glucose plus insulin; beta₂ agonist if there is no ischemic cardiac disease; bicarbonate if there is acidosis.
- Therapy directed at actual removal of potassium from the body: sodium polystyrene sulfonate (Kayexalate®) or dialysis.
- Determine and correct the underlying cause of the hyperkalemia.

Treatment of Chronic Hyperkalemia Caused by Aldosterone Deficiency

Management depends upon the cause of aldosterone deficiency. Patients with primary adrenal failure require the appropriate hormonal replacement. The treatment of patients with the syndrome of hyporeninemic hypoaldosteronism and chronically elevated serum potassium concentrations begins with dietary counseling and a low-potassium diet, because the primary problem is renal retention of potassium secondary to aldosterone deficiency. The hyperkalemia will generally respond to loop diuretics, which provide increased sodium delivery to the collecting tubule resulting in increased potassium excretion, or to the administration of loop diuretics plus the potent mineralocorticoid fludrocortisone. The main side effect of fludrocortisone is renal sodium retention and ECFV overload, so this drug should be started while the patient is under close observation. Careful attention must be given to body weight and the patient must be monitored for signs of ECFV overload such as pedal edema and findings of congestive heart failure. In general, fludrocortisone should be avoided in patients with a significant history of congestive heart failure or other conditions associated with sodium retention. Patients with chronic interstitial disease and tubular unresponsiveness to aldosterone may be managed with low-potassium diets and sometimes cautiously with loop diuretics, but the hyperkalemia will not respond to fludrocortisone.

Treatment of Chronic Hyperkalemia Due to Renal Failure

The management of patients with renal failure begins with dietary counseling and low-potassium diet, because the primary problem is renal retention of potassium secondary to impaired renal excretion. The degree of renal impairment must be severe (GFR <20%) for hyperkalemia to develop on the basis of renal failure alone. A careful review of medicines and a search for "hidden" sources of potassium (see **Figs. 6-1** and **6-2**) are key to the management of the patient with renal disease. The hyperkalemia may respond to loop diuretics (increased sodium delivery to the collecting tubule), but this may result in volume depletion and further impairment of the GFR. Cation exchange resins may be useful in temporarily controlling chronic hyperkalemia, but ultimately dialysis will be required as the GFR falls to the 10% range and ECFV overload supervenes.

Exercises

1. A 40-year-old man has a serum potassium concentration of 6.5 mEq/L. The lab helpfully reports to you that the blood specimen appeared he-

molyzed prior to analysis. He is receiving IV fluids: D5W with 10 mEq/L KCl at KVO ("Keep Vein Open"). What do you do?

Answer: Stop all administration of potassium (oral, enteral, or IV). Even though we suspect spurious hyperkalemia, we must follow an algorithm and take immediate action because this degree of hyperkalemia is potentially life-threatening.

- Obtain a stat ECG. Peaked T waves confirm that true hyperkalemia is present. More advanced changes dictate that urgent treatment to counteract the cardiac effects of hyperkalemia is necessary.
- Quickly make a mental inventory of possible "hidden" sources of potassium and causes of hyperkalemia:

    Potassium penicillin
    Salt substitutes (many contain KCl)
    Hemolysis
    Blood transfusion
    Gastrointestinal hemorrhage (GI absorption of potassium)
    Rhabdomyolysis
    Burns
    Major surgery
- Stop medications that can cause hyperkalemia
- Send stat repeat potassium (drawn without tourniquet if possible to avoid hemolysis).

2. A 36-year-old man presents with diabetic ketoacidosis, arterial blood pH 7.10, and a serum potassium concentration 5.0 mEq/L. The serum creatinine is normal. The ECG is normal except for sinus tachycardia at 108/min. What would you do about the potassium at this point?

Answer: A normal potassium concentration in diabetic ketoacidosis implies significant potassium deficiency in this situation. If there is adequate renal function and urine volume, potassium replacement should be started along with insulin and saline. Remember that insulin will transport potassium into cells. The potassium concentration must be followed very closely to avoid serious hypokalemia. *The major point is, that in diabetic ketoacidosis, a high-normal potassium concentration actually reflects potassium deficiency.* Patients with diabetic ketoacidosis may also have deficiencies of magnesium and phosphate.

3. A 60-year-old 60 kg woman presents with malaise, nausea, decreased appetite, and the new onset of itching. She is a chronic renal failure patient with a creatinine of 7.4 mg/dl and a creatinine clearance that is about 10% of normal. Her data include: sodium 143 mEq/L, potassium 6.6 mEq/L, chloride 107 mEq/L, bicarbonate 16 mEq/L, Hgb. 10.1, Hct 31, WBC 3400. Arterial blood gas: pH 7.32, P_{CO2} 32. What is the cause of the hyperkalemia, and what would your management be?

Answer:

- Obtain a stat ECG. Peaked T waves confirm that true hyperkalemia is present. More advanced changes require emergent therapy to counteract the cardiac effects of hyperkalemia.
- Quickly make a mental inventory of possible "hidden" sources of potassium and causes of hyperkalemia.
- Send stat repeat potassium (drawn without tourniquet if possible to avoid hemolysis). In this case, the impaired GFR has resulted in sufficiently impaired renal potassium excretion that this patient no longer has sufficient GFR to "handle" the daily potassium load from the diet.
- Dietary instruction to avoid foods rich in potassium
- Careful review of all medications
- Treatment of the metabolic acidosis with oral sodium citrate. Citrate is converted to bicarbonate by the liver
- Potassium exchange resin sodium polystyrene sulfonate (Kayexalate®) in sorbitol to correct the acute hyperkalemia

4. You are called to see a 46-year-old diabetic woman with a serum potassium concentration of 6.3 mEq/L. She is asymptomatic. What is your approach?

Answer: Stop any administration of potassium (oral, enteral, or IV). The combination of diabetes mellitus and hyperkalemia should immediately bring to mind the syndrome of hyporeninemic hypoaldosteronism (SHH), but we stick to the algorithm because hyperkalemia is potentially life-threatening and we do not want to overlook anything.

- Obtain a stat ECG. Peaked T waves confirm that true hyperkalemia is present. More advanced changes dictate that urgent treatment to counteract the cardiac effects of hyperkalemia is necessary.
- Quickly make a mental inventory of possible "hidden" sources of potassium and causes of hyperkalemia.
- Send stat repeat potassium (drawn without tourniquet if possible to reduce the risk of hemolysis).

The remainder of the electrolytes and the serum creatinine will be helpful. If the creatinine is normal, we may exclude renal failure as the cause of the hyperkalemia. The patient may have a mild normal anion gap metabolic acidosis. This would support a diagnosis of SHH. Patients with SHH may have mild to moderate renal insufficiency with GFR reduced to 50–75% of normal, but not severe enough (10% of normal) for hyperkalemia to develop on the basis of renal insufficiency alone. The management of these patients begins with dietary counseling and a low-potassium diet, because the primary problem is renal retention of potassium secondary to aldosterone deficiency. The hyperkalemia will generally respond to loop diuretics (increased sodium delivery to the collecting tubule) or to loop diuretics plus the potent mineralocorticoid fludrocortisone. The main side effect of fludrocortisone is renal sodium retention and ECFV overload, so this medication should be started with the patient under close observation. Careful attention must be given to body weight and the patient observed for signs of ECFV overload such as pedal edema and signs of congestive heart failure. In general, fludrocortisone should be avoided in patients with a history of congestive heart failure or other conditions associated with sodium retention. Primary adrenal failure, tubular unresponsiveness to aldosterone, and drugs that interfere with aldosterone should also be considered in the differential diagnosis.

5. Review question. A 79-year-old man (60 kg body weight) with a history of multi-infarct dementia is bedridden and requires enteral tube feedings. He is found to be tachypneic and poorly arousable, and the nurse tells you that he has been having profuse diarrhea. The following data are obtained: sodium 173 mEq/L, potassium 2.8 mEq/L, bicarbonate 18 mEq/L, chloride 137 mEq/L. The arterial blood gas: pH 7.22, P_{CO2} 45, bicarbonate 18 mEq/L. The urine volume is .6 L/24 hours with an osmolality of 670 mOsm/L and sodium of 8 mEq/L.

What is causing the *hypernatremia*?

Answer: Tube feedings can cause diarrhea, which leads to hypernatremia from loss of water in stool, hypokalemia from loss of potassium in stool, and metabolic acidosis from loss of bicarbonate in stool. The free water deficit is approximately:

$$H_2O \text{ deficit} = TBW \times ([Na^+_{(measured)}] - [Na^+_{(normal)}])/[Na^+_{(normal)}]$$

where TBW is total body water, $[Na^+_{(measured)}]$ is the measured serum sodium concentration, and $[Na^+_{(normal)}]$ is the normal serum sodium concentration.

$$H_2O \text{ deficit} = TBW \times ([Na^+_{(measured)}] - [Na^+_{(normal)}])/[Na^+_{(normal)}]$$
$$= .6 \times 60 \times (173-140)/140 = 8.5 \text{ L}$$

or, using TBW = .5 × weight (kg) because this is an elderly patient:

$$= .5 \times 60 \times (173-140)/140 = 7.1 \text{ L}$$

So the total water deficit is probably somewhere between 7.1 L and 8.5 L. This formula can also give the amount of water to be given in order to reduce the sodium concentration to a desired value. Suppose we wanted to correct him to a sodium concentration of 163 mEq/L in the first 12 hours, and assuming 0.5 L urine volume and 0.5 L insensible loss (the patient is tachypneic) in this 12 hour period:

$$H_2O \text{ to bring sodium to } 163 \text{ mEq/L}$$
$$= TBW \times ([Na^+_{(measured)}] - [Na^+_{(desired)}])/[Na^+_{(desired)}]$$
$$= .5 \times 60 \times (173-163)/160 = 1.8 \text{ L}$$

Now, add 0.5 L for insensible losses. The volume of water (which could be given as D5W) over the next 12 hours would be about 2.3 L. Remember to keep an eye on continued losses: If the patient continues to lose water in stool, then the amount of water needed to correct the patient to 163 mEq/L will be greater. The sodium concentration should be checked every 2–3 hours during treatment to avoid overly rapid correction, or inadequate correction. The formula will give only an *approximation* of the actual water requirement.

What about the *hypokalemia*?

Answer: The potassium deficit in this patient is very severe, in view of the pH 7.22. Remember that metabolic acidosis shifts potassium out of cells leading to the misleadingly *high* concentration of 2.8 mEq/L in this case. This tells us that a severe potassium deficit is present. Severe hypokalemia in the setting of significant metabolic acidosis is a medical emergency. This man should be given potassium replacement in a closely monitored setting. I would *not* treat this man with bicarbonate initially, even though he is acidemic. The resultant increase in pH could worsen the hypokalemia and precipitate cardiac arrhythmias. After potassium replacement is well underway, I could reevaluate the need for bicarbonate. The other reason I would not give bicarbonate initially is that part of the drop in pH is explained by a *respiratory acidosis* in addition to the moderate metabolic acidosis caused by the diarrhea. Mixed acid-base disorders are covered in Chapter 9. Try this example again after working through Chapter 9.

CHAPTER 7. METABOLIC ACIDOSIS

Causes of Metabolic Acidosis

Type I = ↑ H⁺

Metabolic acidosis is a process that causes a primary decrease in the plasma HCO_3^- concentration. Metabolic acidosis is generated by either a gain of acid or a loss of HCO_3^-. Gain of acid may result from

- Increased endogenous hydrogen ion production, as in ketoacidosis, L-lactic acidosis, D-lactic acidosis and salicylate intoxication.
- Metabolism of ingested toxins such as methanol, ethylene glycol, and paraldehyde.
- Decreased renal excretion of hydrogen ion as in uremic acidosis and distal (type I) renal tubular acidosis.

Loss of HCO_3^- may result from

Type II = loss of HCO_3^-

- Renal loss in proximal (type II) renal tubular acidosis.
- Gastrointestinal loss in diarrhea.

Typically, metabolic acidosis is classified according to whether or not there is an increase in the anion gap (see **Fig. 7-1**).

High Anion Gap Metabolic Acidosis

A high anion gap acidosis results from either the production of an endogenous acid (ketoacidosis, lactic acidosis, uremic acidosis, salicylate intoxication) or the addition of certain exogenous compounds (ethylene glycol, and methanol). A high anion gap acidosis is caused by the addition of a hydrogen ion plus an unmeasured anion. The H^+ is buffered by HCO_3^-, and therefore the HCO_3^- concentration falls. The unmeasured anion increases the term UA and therefore increases the anion gap according to:

$$AG = UA - UC = [Na^+] - ([Cl^-] + [HCO_3^-])$$

97

FIGURE 7-1. Common Causes of Metabolic Acidosis

Increased anion gap
  Diabetic ketoacidosis
  L-lactic acidosis
  D-lactic acidosis
  Alcoholic ketoacidosis
  Uremic acidosis (advanced renal failure)
  Salicylate intoxication
  Ethylene glycol intoxication
  Methanol intoxication
  Paraldehyde intoxication
Normal anion gap
  Mild to moderate renal failure
  Gastrointestinal loss of HCO_3^- (acute diarrhea)
  Type I (distal) renal tubular acidosis
  Type II (proximal) renal tubular acidosis
  Type IV renal tubular acidosis
  Dilutional acidosis
  Treatment of diabetic ketoacidosis (ketones lost in urine)

Diabetic ketoacidosis (DKA)

Patients with severe diabetic ketoacidosis typically present with

- High anion gap metabolic acidosis
- Severe acidemia (pH < 7.15)
- Hyperglycemia
- ECFV depletion
- Potassium depletion despite serum potassium concentrations that may be normal or elevated.

In DKA, the anion gap acidosis is due to the generation of ketoacids, which are produced by the incomplete oxidation of fatty acids. Typically, diagnosis of diabetic ketoacidosis is made in the setting of uncontrolled insulin-dependent diabetes mellitus with tachypnea, polyuria, polydipsia, severe acidemia, and a urine dipstick positive for ketones. Serum ketones are generally strongly positive in diabetic ketoacidosis. Occasionally in diabetic ketoacidosis, a dipstick test for ketones may underestimate the degree of ketosis because of a marked increase in the beta-hydroxy butyrate/acetoacetate ratio. This is because the test reagents of the dipstick do not detect beta-hydroxy butyrate.

L-lactic acidosis

L-lactic acidosis is by far the most common form of lactic acidosis. The most common cause of L-lactic acidosis is tissue hypoxia. L-lactic acidosis is typically divided into those disorders associated with hypotension or severe hypoxemia (Type A) and all other causes (Type B). The latter can be further divided into

- Common conditions including sepsis, severe liver disease, diabetes mellitus, and various malignancies
- Lactic acidosis caused by toxins or medications, including phenformin and ethanol
- Rare hereditary forms

D-lactic acidosis

D-lactic acidosis is a very uncommon metabolic disturbance. It occurs in patients with a short-bowel, typically several months or a few years after a small bowel bypass created for the treatment of refractory massive obesity. Patients with D-lactic acidosis have episodes of neurological dysfunction characterized by ataxia, slurred speech, and confusion, in association with a high anion gap metabolic acidosis.

The acidosis is caused by fermentation of incompletely digested carbohydrate by anaerobic bacteria, resulting in the production of D-lactic acid which is poorly metabolized in animals. The preliminary diagnosis of D-lactic acidosis must be based on the clinical presentation, because routine clinical laboratories only detect L-lactate, which will be normal.

Alcoholic ketoacidosis

Alcoholic ketoacidosis is a common, serious condition that occurs in some chronic alcoholics, especially after prolonged binge drinking with diminished caloric intake. Accelerated ketogenesis results from the combined effects of starvation and ethanol. The patients may have nausea, vomiting, and abdominal pain. Metabolic alkalosis from vomiting and respiratory alkalosis may also be present in addition to the high anion gap acidosis. There is often ECFV depletion, hypoglycemia, and GI bleeding or acute pancreatitis. Because of prolonged poor food intake, phosphorous and magnesium are often depleted, even though the serum concentrations of phosphate and magnesium at presentation may be within the normal range. A dipstick test for urine ketones (which does not detect beta-hydroxy butyrate) may underestimate the degree of ketosis in alcoholic ketoacidosis because of a marked increase in the beta-hydroxy butyrate/acetoacetate ratio.

Uremic acidosis

High anion gap metabolic acidosis does not usually occur in renal failure until the GFR declines below 20% of normal. The patient with mild to moderate chronic renal failure frequently has a normal anion gap acidosis rather than a high anion gap acidosis. The normal anion gap acidosis results from failure of renal elimination of hydrogen ion by production and excretion of NH_4^+.

As the GFR falls, there is an increase in the anion gap due to retention of un-measured anions such as sulfate, phosphate, and organic anions.

Salicylate intoxication

Although salicylate intoxication may produce a high anion gap acidosis, the most common and earliest manifestation is respiratory alkalosis due to the effect of salicylate on the respiratory center. The metabolic acidosis, which may develop after the respiratory alkalosis appears, is caused by the salicylate interfering with certain metabolic processes. This interference leads to the increased accumulation of organic acids, such as lactic acid and ketoacids, which cause the acidosis and the increase in the anion gap. The salicylate itself makes up only a small part of the increased anion gap.

Ethylene glycol intoxication

Ethylene glycol intoxication may result from drinking antifreeze or radiator fluid. There is usually, but not always, a background of chronic alcoholism. Ethylene glycol is metabolized by alcohol dehydrogenase to a number of very toxic compounds which produce:

- A high anion gap acidosis.
- Acute central nervous system dysfunction: ataxia, confusion, seizures, and coma
- Acute renal failure
- Calcium oxalate crystals in the urine (one of the toxic compounds is oxalic acid)

Methanol intoxication

Methanol is metabolized to formic acid by alcohol dehydrogenase. Formic acid contributes to the high anion gap acidosis seen in this serious intoxication. Optic neuritis with blindness, and pancreatitis may develop in methanol intoxication.

A few comments about the anion gap

One might suppose that in a high anion gap metabolic acidosis, there would be a correlation between the increase in the anion gap, which is caused by the addition of the anion to the ECF, and the decrease in the bicarbonate, which is caused by the titration of HCO_3^- by the hydrogen ion. According to the equation

$$AG = [Na^+] - ([Cl^-] + [HCO_3^-])$$

one could logically expect that if the AG increases because of a high anion gap acidosis, the HCO_3^- concentration would decrease by an equal amount. For example, if a lactic acidosis or diabetic ketoacidosis increases the anion gap by 15 mEq/L, the HCO_3^- concentration might be expected to fall by an equal amount, 15 mEq/L.

A one-to-one relationship between the increase in the anion gap and the decrease in bicarbonate is often *not* the case, however. One reason is that hydrogen ion is buffered intracellularly and by bone as well as by the HCO_3^- in extracellular fluid. Simply put: HCO_3^- does not have to buffer all the hydrogen ion by itself, but "gets help" from other buffer systems. Therefore, the $[HCO_3^-]$ may decrease by an amount *less* than the increase in the anion gap. For lactic acidosis, the ratio of the increase in the AG to the decrease in the $[HCO_3^-]$ is not usually 1.0, but on the average may actually be closer to 1.5 because of this extra buffering of hydrogen ion outside the ECF. That is, for lactic acidosis, approximately:

Change in AG/Change in $[HCO_3^-]$ = 1.5

Lactic Acid
$$\Delta [HCO_3] = \frac{\Delta AG}{1.5}$$

or, rearranging:

Change in $[HCO_3^-]$ = Change in AG/1.5

∴ lactic acidosis more likely to be absorbed by bone/cell.

Using this very rough formulation, we might expect that if a lactic acidosis increases the AG by 15 mEq/L, then the $[HCO_3^-]$ would fall by about: Change in AG/1.5 = 15/1.5 = 10 mEq/L, not 15 mEq/L.

For ketoacidosis, the ratio of the increase in the AG to the decrease in the $[HCO_3^-]$ *is* closer to 1.0, perhaps because some ketoanions, which constitute the increase in the AG, may be lost in the urine. Therefore, for ketoacidosis, approximately:

Ketoacidosis

Change in $[HCO_3^-]$ = Change in AG

It should be carefully restated that this is a *very rough* way to estimate the expected fall in $[HCO_3^-]$ for a given increase in AG when there is a lactic acidosis or a ketoacidosis. For uremic acidosis and the other causes of high anion gap metabolic acidosis, the relationship between the increase in the AG and the decrease in the bicarbonate is unpredictable.

How can we use this information in the setting of lactic acidosis or ketoacidosis? A measured $[HCO_3^-]$ much higher than predicted by the increase in anion gap is a clue that a "hidden" metabolic alkalosis may also be present. A measured $[HCO_3^-]$ much less than predicted by the increase in anion gap is a clue that a "hidden" normal anion gap metabolic acidosis may also be present.

When I diagnose a high anion gap acidosis due to a lactic acidosis or a ketoacidosis, I compare the predicted fall in bicarbonate (based upon the increase in anion gap) to the actual fall in bicarbonate, then use the following guidelines:

- A measured [HCO_3^-] much higher than predicted by the increase in anion gap is a clue that a "hidden" metabolic alkalosis may also be present.
- A measured [HCO_3^-] much less than predicted by the increase in anion gap is a clue that a "hidden" normal anion gap metabolic acidosis may also be present.

This is a preliminary discussion on the use of the anion gap. If this section is not clear at this point, don't worry. The use of the anion gap will be dealt with more thoroughly in Chapter 9.

Normal Anion Gap Metabolic Acidosis

The normal anion gap metabolic acidoses are discussed in terms of whether they are usually associated with hypokalemia or hyperkalemia.

Mild to moderate renal failure

The patient with mild to moderate chronic renal failure will generally have a normal anion gap acidosis rather than a high anion gap acidosis because failure of renal ammoniagenesis impairs the ability of the kidney to rid the body of excess hydrogen ions. The patient will often have a normal serum potassium concentration, but hyperkalemia may develop in the presence of a potassium load or if more severe renal failure develops (when the GFR falls below 10–20%).

Gastrointestinal loss of HCO_3^- (usually hypokalemic)

Acute secretory diarrhea often results in a normal anion gap metabolic acidosis with hypokalemia. If the metabolic acidosis is severe and losses of alkali in the stool continue, HCO_3^- replacement may be warranted.

Distal (Type I) renal tubular acidosis (usually hypokalemic)

Distal renal tubular acidosis (RTA) is due to inability of the renal tubule to eliminate hydrogen ion adequately. The clinical picture consists of metabolic acidosis, inability to lower the urine pH below 5.3 despite the presence of acidemia, and hypokalemia with renal potassium loss. The serum potassium may be in the range of 2.0–3.0 mEq/L or less. In addition, these patients may have calcium phosphate stones and nephrocalcinosis. Renal potassium loss generally corrects after treatment of ECFV depletion and acidosis.

There are many causes of distal RTA. Most of them are uncommon (see **Fig. 7-2**).

Proximal (Type II) RTA (usually hypokalemic or normokalemic)

The syndrome of metabolic acidosis caused by impaired proximal tubular HCO_3^- reabsorption is called proximal RTA. When a defect develops in

FIGURE 7-2. Causes of Renal Tubular Acidosis

(handwritten) Type I = ↑H+

Type I (Distal)
Hereditary
Acquired
Hyperparathyroidism
Sjogren's syndrome
Medullary sponge kidney
Amphotericin B
Chronic tubulointerstitial disease

Type II (Proximal)

(handwritten) Type II = ↓⁻HCO₃

Hereditary
Acquired
Multiple myeloma
Mercury
Lead
Acetazolamide
Wilson's disease

Type IV (hyperkalemic)
Aldosterone deficiency
Primary adrenal failure
Syndrome of hyporeninemic hypoaldosteronism (SHH)
Chronic interstitial nephritis
Analgesic nephropathy
Chronic pyelonephritis
Obstructive nephropathy
Sickle cell disease
Drugs
Amiloride
Spironolactone
Triamterene
Trimethoprim
Heparin therapy
Nonsteroidal antiinflammatory drugs
Angiotensin-converting enzyme inhibitors

the proximal tubular reabsorption of filtered HCO_3^-, the HCO_3^- concentration falls as HCO_3^- is lost in the urine. This lowering of the HCO_3^- concentration results in metabolic acidosis. The abnormality in proximal RTA is sometimes viewed as a resetting of the "threshold" for bicarbonate reabsorption by the proximal tubule to a lower value. Type II RTA is sometimes associated with defective proximal tubular reabsorption of several other solutes, including amino acids, glucose, phosphorous, and urate. This generalized failure of proximal tubular reabsorption is called the Fanconi syndrome. Patients with proximal RTA do not have a problem secreting hydrogen ion. Therefore, patients with proximal RTA *can* acidify their urine to a pH of less than 5.3.

There are many causes of proximal RTA. Most of them are uncommon (see **Fig. 7-2**).

(handwritten) Faconi Syndrome → generallized failure of proximal RTA

Type IV RTA (usually hyperkalemic)

This disorder has been discussed in Chapter 6, under the discussion of hyperkalemia secondary to aldosterone deficiency or tubular unresponsiveness to aldosterone. Common causes of type IV RTA are listed in **Fig. 7-2**. This RTA, due to either aldosterone deficiency or tubular unresponsiveness to aldosterone, results in a normal anion gap metabolic acidosis and hyperkalemia. One common cause of type IV RTA results from hyporeninemic hypoaldosteronism, which is characterized by low plasma renin activity, diminished plasma aldosterone, and hyperkalemia. This syndrome is seen in a variety of renal disorders, but the most common cause is diabetes.

Tubular unresponsiveness to aldosterone occurs with a number of chronic renal disorders. The syndrome is quite similar to hypoaldosteronism, but these patients do not respond to mineralocorticoid replacement.

Dilutional acidosis

Mild acidosis may result from the dilution of plasma HCO_3^- by rapid extracellular volume expansion with large amounts of fluid lacking in bicarbonate or bicarbonate precursors such as lactate. This cause of acidosis may be seen in critical care units following massive fluid resuscitation.

Respiratory Compensation for Metabolic Acidosis

The hydrogen ion concentration of ECF is determined by the ratio of the P_{CO_2} (which is controlled by the lungs) to the $[HCO_3^-]$ (which is controlled by the kidneys) according to the relation:

$$[H^+] \propto P_{CO_2}/[HCO_3^-]$$

A metabolic acidosis is a process that causes a primary decrease in $[HCO_3^-]$. The respiratory compensation for a metabolic acidosis is increased ventilation, which produces a secondary decrease in P_{CO_2}. This returns the $P_{CO_2}/[HCO_3^-]$ ratio, and therefore the hydrogen ion concentration, toward the normal range. Typically, the lungs do not return the hydrogen ion concentration all the way into the normal range, but only toward the normal range. What should the P_{CO_2} be after compensation for a metabolic acidosis? The quantitative answer to this question is obtained by using the formula for expected respiratory compensation for a metabolic acidosis. That is, the P_{CO_2} should be equal to:

$$P_{CO_2} = 1.5 \times [HCO_3^-] + 8$$

What if the measured P_{CO_2} differs from this predicted value? A significant difference means that there is also a coexisting respiratory disorder in addition to the metabolic acidosis, because the P_{CO_2} is not behaving as we would expect. If the measured P_{CO2} is higher than predicted by the formula, there is a coexisting respiratory acidosis. If the measured P_{CO2} is lower than predicted, there is a coexisting respiratory alkalosis. This formula is approximate, and we should allow the measured P_{CO2} to be ± 2 mm Hg off from that predicted by the formula. A significant deviation from the value predicted by the formula in either direction, however, indicates that in addition to a metabolic acidosis, there is also a coexisting respiratory disorder.

Organic → NaHCO₃ only in < 7.2 pH.

Treatment of Metabolic Acidosis   Non AG → bicarb

Measuring the pH is important, because when arterial blood pH drops below 7.20 there may be impaired cardiac contractility. The etiology of the acidosis is also important. In patients with metabolic acidosis produced by organic acids (for example ketoacidosis, lactic acidosis), treatment with bicarbonate is indicated only for severe acidemia (pH <7.10; bicarbonate <10 mEq/L). This is because metabolism of the organic anion (ketoacids, lactate) yields bicarbonate, and administration of alkali to such a patient may induce "overshoot" metabolic alkalosis. There are other potential adverse effects of administration of $NaHCO_3$ in situations where the acidemia is not life-threatening, including iatrogenic hypernatremia, volume overload, increased intracellular acidosis, and cardiovascular compromise.

On the other hand, patients with metabolic acidosis produced by the non anion gap acidoses, methanol, ethylene glycol, and chronic renal failure may require bicarbonate administration with less severe acidemia. **Fig. 7-3** gives a general approach to the treatment of metabolic acidosis, but the details of treatment depend upon the specific cause and severity of the metabolic acidosis.

Treatment of Diabetic Ketoacidosis

Diabetic ketoacidosis generally responds well to therapy with insulin, saline, and potassium. This therapy corrects the hyperglycemia, ketogenesis, acidemia, and potassium deficit. Because the circulating ketoanions will be converted to HCO_3^- by the liver once insulin and fluids reverse ketosis, they represent "potential" HCO_3^-. The pH and the HCO_3^- concentration increase as the anion gap decreases with appropriate therapy of diabetic ketoacidosis. The majority of patients should not receive HCO_3^- replacement for this reason.

There are some important points to remember in the management of diabetic ketoacidosis:

• Despite potassium deficiency, the initial potassium concentration may be misleadingly normal or even high because of insulin deficiency and hypertonicity. Therefore, a normal or even elevated serum potassium may disguise

FIGURE 7-3. General Guidelines in Treatment of Metabolic Acidosis

1. Early intervention should always be directed at identifying and correcting the specific underlying cause(s) of the metabolic acidosis.
2. In general, some types of metabolic acidosis will require HCO_3^- therapy and some will not. Therefore, determining the cause of the metabolic acidosis is central to appropriate management.
3. The decision to use HCO_3^- replacement should be weighed carefully, depending upon the severity of the acidemia (blood pH) and the type of acidosis.
4. If the pH falls below 7.10, emergency HCO_3^- administration should be considered, regardless of the cause of the acidosis. This is especially important if there appears to be respiratory fatigue or developing hemodynamic instability.
5. Never give HCO_3^- without a determination of blood pH: An offsetting metabolic or respiratory alkalosis may be moving the pH up, in which case HCO_3^- administration could result in life-threatening alkalemia.
6. When the decision to give IV bicarbonate is made in the acute setting, calculate the amount of HCO_3^- required to increase the HCO_3^- concentration to a specified value, often several mEq/L above the measured value. In general, assume that HCO_3^- distributes in about 50% of Body weight (kg).

$$HCO_3^- \text{ deficit} = .5 \times \text{Body weight (kg)} \times ([HCO_3^-_{(desired)}] - [HCO_3^-_{(measured)}])$$

7. In severe acidosis (pH in the 7.10 range, HCO_3^- <10 mEq/L), the amount of HCO_3^- required to increase the HCO_3^- concentration to the range 10-12 mEq/L is initially calculated. For example, in a 70 kg patient, if the HCO_3^- is 6 mEq/L, and it is desired to bring the HCO_3^- to 10 mEq/L:

$$HCO_3^- \text{ deficit} = .5 \times 70 \times (10 \text{ mEq/L} - 6 \text{ mEq/L}) = 140 \text{ mEq/L}$$

Give this calculated amount slowly and remeasure pH, HCO_3^- and P_{CO2} after the HCO_3^- is given to assess the effect of therapy on the acid-base status.
8. The relationship between amount of HCO_3^- given and the increase in HCO_3^- is *not* linear: At mild levels of acidemia, 2 mEq/kg will increase the HCO_3^- by roughly 4 mEq/L. At severe levels of acidemia, 2 mEq/kg will only raise the HCO_3^- by roughly 2 mEq/L.
9. In the case of an ongoing acidosis, repeated doses of HCO_3^- may be required until the underlying cause of the acidosis can be corrected.

---

a severe depletion of total body potassium. A normal potassium concentration in diabetic ketoacidosis suggests significant potassium deficiency.

- Therapy of diabetic ketoacidosis tends to rapidly lower the serum potassium concentration because potassium enters cells. This is because pH rises with appropriate therapy, insulin is administered, and hypertonicity is corrected. The potassium concentration may plummet, leading to cardiac arrhythmias. It is advisable to start potassium replacement as soon as it is established that the patient is not anuric and that the ECG does not show advanced signs of hyperkalemia.
- Some patients may develop a normal anion gap acidosis during the course of treatment due to the loss of ketoanions (which are potential HCO_3^-) in the urine. In these patients, the anion gap decreases during treatment, but the pH and the HCO_3^- concentration do not increase as expected. If the resulting

normal anion gap metabolic acidosis is severe, these patients may require HCO_3^- replacement.

Treatment of L-Lactic Acidosis

The physician should first treat the underlying disorder (shock, sepsis, etc.). Administration of alkali does not reverse the underlying cause of lactic acidosis, but when the pH is less than 7.10 it will help to protect against the dangerous effects of severe acidemia. Consider alkali therapy in cases of severe lactic acidosis when the pH falls below 7.10. When the underlying condition is corrected, however, lactate is converted to HCO_3^-, and there may be an "overshoot alkalosis" during recovery. There are other potential adverse effects of administration of HCO_3^- in situations where the acidemia is not life-threatening, including metabolic alkalosis, hypernatremia, volume overload, and cardiovascular compromise.

Treatment of D-Lactic Acidosis

Treatment of acute D-lactic acidosis may include intravenous fluids and HCO_3^-, but also requires oral antibiotics to eliminate the offending flora. Chronic therapy involves oral antibiotics and either dietary restriction of carbohydrate, temporary fasting with hyperalimentation, or taking down of the small bowel bypass.

Treatment of Alcoholic Acidosis

Treatment consists of the administration of dextrose-containing saline to reverse ketogenesis and correct any ECFV depletion. D5 0.9% saline with supplemental KCl is usually appropriate for this purpose. HCO_3^- is not usually required because the ketones are converted to HCO_3^-, once the ketosis is reversed and the ECFV normalized. In general, patients will require substantial potassium replacement, which should begin promptly. In the case of alcoholic ketoacidosis with severe hypokalemia, the administration of glucose should be postponed until potassium replacement is well underway because glucose can stimulate insulin release, which can acutely worsen the hypokalemia.

A word of caution: *It is dangerous to give dextrose to chronic alcoholics or other patients who are malnourished, without giving thiamine first.* Glucose administration without thiamine sometimes precipitates acute Wernicke's encephalopathy in a chronic alcoholic and can lead to permanent neurological impairment. We generally give a "rally pack" to these patients, consisting of:

- 100 mg thiamine IM stat
- 5 mg folate added to the IV solution
- 1 ampule of multiple vitamins added to the IV solution
- The thiamine may be given 2–3 times to make sure stores are replete.

We also check for evidence of calcium, phosphorous, and magnesium deficiencies, which may not become apparent for 2–3 days.

Treatment of Salicylate Intoxication

Treatment is directed at increasing excretion of the salicylate. Salicylate is excreted more efficiently in an alkaline urine, so HCO_3^- is sometimes added to the intravenous fluids *unless the pH of blood is alkaline due to a predominant respiratory alkalosis*. It is important to assess the arterial blood gas values first to avoid giving HCO_3^- to a patient with an alkaline pH. If toxicity is severe and neurologic symptoms are present, hemodialysis may be required.

Treatment of Ethylene Glycol Intoxication

Treatment is threefold:

- Administer ethanol to impair the conversion of ethylene glycol to its toxic metabolites.
- Dialyze to remove the ethylene glycol, even before acute renal failure develops.
- HCO_3^- may be required in cases of severe acidosis because the organic anions generated by ethylene glycol are not converted to HCO_3^- as are the ketoanions in diabetic ketoacidosis. That is, the anions do not constitute a source of "potential" HCO_3^-.

Treatment of Methanol Intoxication

Treatment is the same as for ethylene glycol:

- Administer ethanol to impair the conversion of methanol to its toxic metabolites.
- Dialyze to remove the methanol.
- HCO_3^- may be required in cases of severe acidosis because the organic anions generated by methanol are not converted to HCO_3^- and do not constitute a source of "potential" HCO_3^-.

Treatment of Distal (Type I) RTA

Determine and correct the cause, if possible, and replace HCO_3^- and potassium. Distal RTA may require an amount of HCO_3^- replacement that roughly equals the daily production of hydrogen ion (50–100 mEq/day). Some of the HCO_3^- should be given as $KHCO_3$ to correct potassium losses as long as there is no renal failure. Practically, the HCO_3^- is given by mouth as citrate, which is a HCO_3^- precursor and tends to cause less bloating than HCO_3^-.

Treatment of Type II (Proximal) RTA

HCO_3^- can be given as $KHCO_3$ (often as K-citrate) as long as there is no significant degree of renal failure. The abnormality in proximal RTA is resetting the "threshold" for bicarbonate reabsorption by the proximal tubule to a lower value. Consequently, it is difficult to maintain the HCO_3^- concentration much above the threshold concentration, even with large doses of HCO_3^- because the administered HCO_3^- is rapidly lost in the urine as soon as the HCO_3^- concentration is raised above the threshold for HCO_3^- reabsorption. Consequently, the amount of HCO_3^- replacement in PRTA may be quite large.

Mild to moderate hypokalemia is common in PRTA and is worsened by alkali therapy. Renal potassium loss increases when bicarbonate is administered to correct the acidosis because HCO_3^- is not reabsorbed well proximally and travels to the collecting tubule, in association with sodium. Therefore, the administered HCO_3^- acts as a non-reabsorbable anion that increases distal sodium delivery and therefore sodium-potassium exchange and urine potassium loss.

Treatment of Type IV RTA

The management of patients with type IV RTA depends upon the cause of the aldosterone deficiency or tubular unresponsiveness to aldosterone (**Fig. 7-2**). Primary adrenal failure should be treated with the appropriate hormonal replacement. The treatment of patients with the syndrome of hyporeninemic hypoaldosteronism begins with dietary counseling and a low-potassium diet, because the primary problem is renal retention of potassium secondary to aldosterone deficiency. The hyperkalemia will generally respond to the administration of loop diuretics or to the administration of loop diuretics plus the potent mineralocorticoid fludrocortisone. The main side effect of fludrocortisone is renal sodium retention and volume overload, so this drug should be started under close observation with careful attention to body weight and observation for signs of sodium excess. In general, fludrocortisone should be avoided in patients with a history of congestive heart failure or other edematous states.

Drugs that interfere with aldosterone release or aldosterone effect on the collecting tubule should be withdrawn if hyperkalemia develops.

Exercises

1. A 40-year-old man is admitted with shallow, rapid respirations. His serum chemistries are: sodium 142 mEq/L, potassium 3.6 mEq/L, chloride 100 mEq/L, bicarbonate 12 mEq/L. Arterial blood gas: pH 7.28, P_{CO_2} 26, HCO_3^- 12. What is your differential diagnosis?
   Answer: The diagnosis of acid-base disorders requires a systematic approach to identify *all* the disorders present in a given patient. Chapter 9

describes a simple, three-step method to use on every single acid-base problem. Do not worry if you do not yet understand all three steps.

Step 1: Identify one disorder. The pH is low and the bicarbonate is also low. Therefore metabolic acidosis is present.

Step 2: Use the formula to see if the compensation is correct (if the compensation for the metabolic acidosis is not what is predicted from the formula, then a coexisting respiratory disorder is also present). For a metabolic acidosis, the P_{CO_2} should be

$$P_{CO_2} = 1.5 \times [HCO_3^-] + 8$$

$$= (1.5 \times 12) + 8 = 26$$

The measured P_{CO_2} is equal to the P_{CO_2} predicted by the formula for expected compensation. This means that there is appropriate compensation, and no respiratory disorder is present. If the patient's P_{CO_2} were significantly higher than this, then we would diagnose a coexisting respiratory acidosis. If the patient's P_{CO_2} were significantly lower than this, then we would diagnose a coexisting respiratory alkalosis.

Remember to use the values of both P_{CO_2} and $[HCO_3^-]$ from the arterial blood gas rather using the P_{CO_2} from the arterial blood gas and the serum $[HCO_3^-]$ for purposes of determining if compensation is appropriate.

Step 3: Calculate the anion gap:

$$AG = [Na^+] - ([Cl^-] + [HCO_3^-])$$

$$= 142 - (100 + 12) = 30$$

There is a high anion gap acidosis present. Remember: If the anion gap is 30 or more, there is a high anion gap acidosis present, even if the pH is normal.

Remember to use all serum values to calculate the anion gap, rather than using the serum sodium and chloride values with the calculated bicarbonate from the arterial blood gas.

The differential diagnosis of a high anion gap acidosis is shown in **Fig. 7-1.** The next step is to determine which disorders the patient has.

• Is there a history of insulin-dependent diabetes mellitus? Is there a history of alcoholism with a recent binge? A history of alcoholism might suggest not only the possibility of alcoholic ketoacidosis, but also the less common disorders ethylene glycol and methanol poisoning. Is there a history of renal failure or of salicylate ingestion? Is shock or severe hypoxemia present?

• Are there ketones in the urine or in the blood? In general, because ketones are concentrated and excreted in the urine, the presence of ketonuria is very sensitive to detect ketosis, but the test for urine ketones may be strongly positive even though the serum level of ketones is only minimally elevated. A strongly positive test for *serum* ketones confirms the suspicion of ketoacidosis.

- The creatinine concentration (renal failure), osmolal gap (ethylene glycol and methanol cause an increase in both the anion gap and the osmolal gap), and lactate concentration (L-lactic acidosis) may also be helpful depending upon the clinical circumstances.
- Ethylene glycol intoxication may be associated with calcium oxalate crystals in the urine.

2. A 20-year-old woman is admitted with protracted vomiting, lethargy, rapid respiration, tachycardia, and a blood pressure of 150/98. She is a known insulin-dependent diabetic who has not been taking her insulin regularly. Her mother tells you that her blood sugars have been "up and down" during the last several months and that she has not been eating well. Her serum chemistries are: sodium 142, potassium 3.6, chloride 106, bicarbonate 16, glucose 230 mg/dl, BUN 70 mg/dl, creatinine: pending. Arterial blood gas: pH 7.28, P_{CO2} 34, HCO_3^- 16. Urine ketones: moderately positive. What is your differential diagnosis, and what do you do to make a diagnosis?

Answer: You make the diagnosis of diabetic ketoacidosis, and begin treatment with insulin, and 0.9% saline with potassium chloride 40 mEq/L at 250 ml/hr.

After 3 hours of treatment, the patient remains lethargic and is now *short of breath*. The anion gap acidosis is not improved. The blood sugar is 70 mg/dl. Where are you?

Answer: You remember to check serum ketones. They are negative. The lactate level is normal. You notice that the patient has developed bibasilar pulmonary rales, indicating ECFV overload from the saline infusion. There has been no urine output since admission. You stop the saline infusion. The serum creatinine finally comes back from the lab: 11.8 mg/dl. The patient does not have diabetic ketoacidosis: She has end-stage renal disease from diabetic nephropathy and is uremic. The high anion gap acidosis is secondary to uremia, and the positive urine ketones are secondary to protracted vomiting, not diabetic ketoacidosis.

I admit that this case is contrived, but there are several lessons:
- At least *consider* the other causes of an anion gap acidosis, even when things look pretty straightforward.
- A moderate level of *urine* ketones may be present in starvation and in patients who have been vomiting. A high reading for *serum* ketones suggests a high degree of ketonemia and is more suggestive of diabetic ketoacidosis (or alcoholic ketoacidosis).
- Always check the urine output and listen for basilar rales when you are giving large volumes of saline.
- In the late stages of diabetic renal disease, as the kidneys fail, insulin requirements may decrease, leading to difficulty with diabetic control.

The urine test for ketones is very sensitive because ketones are concentrated and excreted in the urine. Starvation and protracted vomiting will give relatively high readings for urine ketones, even though the patient does not have significant ketonemia. A test for serum ketones is helpful in confirming a case of diabetic ketoacidosis: The reading will generally be strongly positive in diabetic ketoacidosis, confirming that an elevated anion gap is due to ketoacidosis, but lower or negative in starvation and protracted vomiting.

3. A 60-year-old alcoholic woman is admitted with rapid respiration, tachycardia, and a blood pressure of 90/60. Her chemistries are: sodium 142 mEq/L, potassium 3.6 mEq/L, chloride 100 mEq/L, bicarbonate 12 mEq/L, glucose 180 mg/dl, BUN 28 mg/dl. Arterial blood gas: pH 7.28, P_{CO2} 26, HCO_3^- 12. What is your differential diagnosis, and what do you do to make a diagnosis?

Answer: The anion gap is 30. Therefore, a high anion gap acidosis is present. The differential diagnosis is in **Fig. 7-1**. Serum lactate is normal. We suspect alcoholic ketoacidosis, and we order serum ketones. It is important to mention that there may not *necessarily* be a high reading for ketones in the serum of a patient with alcoholic ketoacidosis as there usually is in diabetic ketoacidosis. The test for ketones, which does not detect beta-hydroxy butyrate, may underestimate the degree of ketosis in alcoholic ketoacidosis because of a marked increase in the beta-hydroxy butyrate /acetoacetate ratio. This is because the test reagents do not react with beta-hydroxy butyrate. Other less common causes of high AG metabolic acidosis in the setting of alcoholism are ethylene glycol and methanol poisonings.

4. A 50-year-old man is admitted with rapid respiration, tachycardia, and a blood pressure of 90/60. His chemistries are: sodium 142 mEq/L, potassium 3.6 mEq/L, chloride 100 mEq/L, bicarbonate 12 mEq/L, glucose 180 mg/dl, BUN 28 mg/dl. Arterial blood gas: pH 7.28, P_{CO2} 26, HCO_3^- 12. Urinalysis: calcium oxalate crystals. What is your differential diagnosis, and what do you do to make a diagnosis?

Answer: High anion gap acidosis. Differential diagnosis is in **Fig. 7-1**. There is a history of alcoholism. A measured osmolality is 360 mOsm/L and there are calcium oxalate crystals in the urine. The calculated osmolality is:

Calculated osmolality

$$= 2 \times [\text{sodium concentration}] + [\text{glucose concentration}]/18$$

$$+ [\text{Blood Urea Nitrogen}]/2.8$$

$$= 2 \times 142 + 180/18 + 28/2.8 = 304 \text{ mOsm/L}$$

The osmolal gap is:

$$\text{Osmolal gap} = \text{OSM}_{(measured)} - \text{OSM}_{(calculated)}$$

$$= 360 - 304 = 56 \text{ mOsm/L}$$

The osmolal gap is markedly increased. The combination of high AG and high osmolal gap suggests either ethylene glycol or methanol poisoning, although a high osmolal gap can occur in ketoacidosis. The calcium oxalate crystals in the urine suggest ethylene glycol.

5. A 30-year-old woman is admitted with tachycardia and a blood pressure of 90/60. She is unable to provide any information. Her chemistries are: sodium 150 mEq/L, potassium 3.1 mEq/L, chloride 123 mEq/L, bicarbonate 12 mEq/L, glucose 180 mg/dl, BUN 28 mg/dl. Arterial blood gas: pH 7.28, P_{CO2} 26, HCO_3^- 12. What is your differential diagnosis?

   Answer: You first calculate the anion gap:

$$AG = [Na^+] - ([Cl^-] + [HCO_3^-])$$
$$= 150 - (123 + 12) = 15$$

   This patient has hypernatremia and a normal anion gap metabolic acidosis. The causes of normal anion gap metabolic acidosis are listed in **Fig. 7-1.** Further questioning reveals that the patient has recently returned from a trip around the world and has been having abdominal cramps, profuse watery diarrhea, and fever for the past 5 days.

6. A 45-year-old 80 kg man presents with sodium 140 mEq/L, potassium 3.8 mEq/L, chloride 110 mEq/L, bicarbonate 8 mEq/L, glucose 180 mg/dl, BUN 28 mg/dl. Arterial blood gas: pH 7.10, P_{CO2} 20, HCO_3^- 6. The patient is developing respiratory fatigue. Calculate the amount of bicarbonate required to bring the bicarbonate from 6 mEq/L to 10 mEq/L. (**Fig. 7-3**).

   Answer: The amount of bicarbonate required to bring the bicarbonate from 6 mEq/L to 10 mEq/L:

   HCO_3^- deficit
   $$= .5 \times \text{Body weight(kg)} \times ([HCO_3^-_{(desired)}] - [HCO_3^-_{(measured)}])$$
   $$= .5 \times 80 \times (10 - 6) = 160mEq$$

   If the decision is made to replace HCO_3^-, give this calculated amount slowly and remeasure pH, HCO_3^- and pCO_2 to assess the effect of therapy on the acid-base status.

7. Calculate the amount of bicarbonate to be given to a 20-year-old 80 kg woman with diabetic ketoacidosis and the following chemistries: sodium 135 mEq/L, potassium 2.6 mEq/L, chloride 93 mEq/L, bicarbonate 10 mEq/L, glucose 480 mg/dl, BUN 42 mg/dl. Arterial blood gas: pH 7.26, P_{CO2} 23, HCO_3^- 10 mEq/L. High reading for *serum* ketones.

   Answer: This patient should not receive bicarbonate. In diabetic ketoacidosis the ketones account for the increase in the anion gap and represent "potential bicarbonate." This means that once the ketosis is reversed by appropriate treatment with insulin, isotonic saline, and potassium, the liver will be able in effect to convert the ketones into bicarbonate. Bicarbonate should *not* be given in this case because:

   • The ketoanions (which account for the increase in the anion gap) will mostly be in effect converted to bicarbonate by the liver.

- The pH of 7.26 is not life-threatening in this case and will likely return toward normal with appropriate therapy. If the pH drops to the range of 7.10 in diabetic ketoacidosis, emergency HCO_3^- administration should be considered, especially if there appears to be respiratory fatigue or developing hemodynamic instability.
- Administration of bicarbonate will cause potassium to move into cells. The low potassium concentration of 2.6 mEq/L may fall rapidly, possibly resulting in serious cardiac arrhythmias.

8. Suppose you have been treating the diabetic ketoacidosis of the patient in the previous exercise for 6 hours with 0.9% saline, insulin and potassium. You receive the following chemistries: sodium 143 mEq/L, potassium 3.6 mEq/L, chloride 112 mEq/L, bicarbonate 16 mEq/L, glucose 180 mg/dl, BUN 28 mg/dl. Serum ketones: negative. Arterial blood gas: pH 7.32, P_{CO2} 32, HCO_3^- 16. What is going on?

Answer: The anion gap is:

$$AG = [Na^+] - ([Cl^-] + [HCO_3^-])$$
$$= 143 - (112 + 16) = 15$$

The AG on admission was 32 mEq/L. The normal anion gap acidosis has developed because ketoanions (which accounted for the increase in the anion gap and represented potential bicarbonate) have been lost in the urine.

9. A 15-year-old boy is admitted with severe diarrhea and the following laboratory data: sodium 142 mEq/L, potassium 3.6 mEq/L, chloride 115 mEq/L, bicarbonate 12 mEq/L, creatinine 1.1 mg/dl. Urine ketones negative. Arterial blood gas: pH 7.12, P_{CO2} 38, HCO_3^- 12. What acid-base disorders are present?

Answer: There are three steps (described more fully in Chapter 9):

Step 1: Identify a single disorder. The pH and the bicarbonate are low. Therefore, metabolic acidosis is present.

Step 2: See if the compensation is correct. If the compensation for the metabolic acidosis is not what is predicted from the formula, then a respiratory disorder is present. For a metabolic acidosis, the P_{CO2} should be

$$P_{CO2} = 1.5 \times [HCO_3^-] + 8$$
$$= 1.5 \times 12 + 8 = 26$$

The patient's P_{CO2} is 38. This is much higher than predicted by the formula. Therefore, the patient has a respiratory acidosis. The respiratory acidosis represents "tiring out" of the patient's respiration and impairment of his ability to compensate for the metabolic acidosis. It could also be a clue to a coincident pulmonary process. *The rising P_{CO2} is a dangerous sign in metabolic acidosis, because a further increase in the P_{CO2} or decrease in the bicarbonate could lead to a precipitous fall in pH.*

Step 3: Calculate the anion gap:

$$AG = [Na^+] - ([Cl^-] + [HCO_3^-])$$
$$= 142 - (115 + 12) = 15$$

The anion gap is normal. Therefore, a normal anion gap acidosis is present, consistent with severe diarrhea. Key in the management of this patient will be close observation of the respiratory status and careful replacement of bicarbonate.

CHAPTER 8. METABOLIC ALKALOSIS

Metabolic alkalosis is a process that causes a primary increase in the plasma bicarbonate concentration. Metabolic alkalosis is *generated* by either the loss of hydrogen ion or the gain of bicarbonate. Generation of a metabolic alkalosis is not enough to sustain an elevation of the HCO_3^- concentration because the kidney normally has a large capacity to excrete excess bicarbonate. For metabolic alkalosis to be sustained, the elevated HCO_3^- concentration must be *maintained* by abnormal renal retention of HCO_3^-. To summarize: metabolic alkalosis requires two things:

• *Generation* by either the loss of hydrogen ion or the gain of HCO_3^-
• *Maintenance* by abnormal renal retention of HCO_3^-

Finding the reasons for the abnormal renal retention of bicarbonate is the key to correcting metabolic alkalosis. When I see a case of metabolic alkalosis, I ask: Why is the kidney abnormally retaining bicarbonate?

Causes of Metabolic Alkalosis

Metabolic alkalosis is discussed according to the mechanism causing the abnormal renal retention of bicarbonate that maintains the metabolic alkalosis (see **Fig. 8-1**). All of the causes of metabolic alkalosis are generally associated with hypokalemia.

ECFV Depletion Syndrome

ECFV depletion increases renal retention of HCO_3^-, even in the presence of a high serum HCO_3^- concentration. A decrease in ECFV increases the proximal reabsorption of HCO_3^- along with sodium. Because ECFV depletion leads to increased HCO_3^- reabsorption, ECFV depletion is an important factor in sustaining an elevated HCO_3^- concentration in patients with metabolic alkalosis. The increased proximal reabsorption of $NaHCO_3$ sustains the metabolic alkalosis until the ECFV depletion is corrected.

In general, patients with metabolic alkalosis associated with ECFV depletion have low concentrations (<10 mEq/L) of chloride in the urine and are re-

FIGURE 8-1.   Causes of Metabolic Alkalosis

ECFV depletion—chloride depletion syndrome (saline-responsive)
  Vomiting/nasogastric suction
  Diuretic therapy
  Posthypercapnea
  Chronic diarrhea/laxative abuse
Severe potassium depletion from any cause (saline-resistant)
Mineralocorticoid excess syndromes (saline-resistant)
  Primary hyperaldosteronism
  Cushing's syndrome
  Ectopic ACTH
  Secondary hyperaldosteronism
    Renovascular disease
    Malignant hypertension
    Congestive heart failure (with diuretic therapy)
    Cirrhosis (with diuretic therapy)
Gitelman's syndrome (saline-resistant)
Bartter's syndrome (saline-resistant)
Metabolic alkalosis maintained by renal failure (saline generally contraindicated)

---

sponsive to NaCl administration as 0.9% saline. The metabolic alkalosis associated with ECFV depletion is sometimes called saline-responsive metabolic alkalosis or chloride-responsive metabolic alkalosis. Patients with the other types of metabolic alkalosis generally have higher concentrations (>20 mEq/L) of chloride in the urine and do not generally respond to saline replacement. The other metabolic alkaloses listed in **Fig. 8-1** are sometimes called saline-resistant metabolic alkaloses or chloride-resistant metabolic alkaloses.

Vomiting/nasogastric suction. Acute vomiting and nasogastric suction account for a majority of the cases of profound metabolic alkalosis. Loss of gastric HCl is the mechanism of *generation* of metabolic alkalosis: The loss of hydrochloric acid results directly in an increase in plasma HCO_3^- concentration, because the loss of one hydrogen ion has the *same* result as the gain of one HCO_3^-. The net result is the addition of one free HCO_3^- to the body. The alkalosis is *maintained* by ECFV depletion.

Diuretic therapy. Administration of thiazide or loop diuretics is a common cause of metabolic alkalosis associated with ECFV depletion and potassium depletion.

Posthypercapnic metabolic alkalosis. Alkalosis occurs in chronically hypercapnic patients who undergo therapy to reduce arterial P_{CO2} acutely, often by mechanical hyperventilation during an episode of acute respiratory failure. The metabolic alkalosis is then maintained by ECFV depletion. Diuretics are often prescribed for patients with respiratory failure, which can lead to worsening of the ECFV depletion and metabolic alkalosis.

Chronic diarrhea. In ulcerative colitis, Crohn's disease of the colon, and chronic laxative abuse, mild to moderate metabolic alkalosis may occur. The alkalosis is maintained by ECFV depletion. Acute secretory diarrhea, on the other hand, is often associated with metabolic acidosis due to stool loss of bicarbonate.

Severe Potassium Depletion

Profound potassium depletion to <2.0 mEq/L, regardless of cause, can maintain a saline-resistant metabolic alkalosis by complex mechanisms. Potassium must be replaced before metabolic alkalosis can be corrected.

Mineralocorticoid Excess

A variety of clinical syndromes are associated with increased mineralocorticoid effect, saline-resistant metabolic alkalosis, and hypokalemia (see **Fig. 8-1**). In general, these disorders each result in an alkalosis that is maintained by high levels of mineralocorticoid activity. The alkalosis is resistant to the administration of normal saline. Mineralocorticoid has the effect of increasing collecting tubule sodium-potassium exchange and increasing collecting tubule hydrogen ion secretion and renal ammoniagenesis, which result in hypokalemia and metabolic alkalosis.

Gitelman's Syndrome and Bartter's Syndrome

Gitelman's syndrome is a very rare entity that is caused by a defect in sodium reabsorption in the sodium-chloride cotransporter in the distal convoluted tubule (the segment of the tubule where thiazide diuretics work). Gitelman's syndrome is characterized by:

• Normotension
• Hypokalemia with renal potassium loss
• Saline-resistant metabolic alkalosis
• High urinary chloride concentration
• High levels of renin and aldosterone
• Hypocalciuria

Gitelman's syndrome is considered separately because several more common disorders (surreptitious vomiting, laxative abuse, and diuretic abuse) are often confused with Gitelman's syndrome. Surreptitious vomiting tends to be a chronic disorder that may present with potassium depletion and metabolic alkalosis. Surreptitious vomiting may mimic Gitelman's syndrome because of its covert pattern, associated electrolyte abnormalities, and secondary increases in plasma renin activity and aldosterone. Covert diuretic abuse or laxative abuse may also be confused with Gitelman's syndrome.

Bartter's syndrome is a very rare entity that generally occurs in neonates or very young children. Bartter's syndrome is due to a defect in sodium chloride reabsorption in the ascending limb of the Loop of Henle where 20–30% of filtered sodium is reabsorbed (the segment of the tubule where *loop* diuretics work). Patients may have marked volume depletion, hypokalemia and metabolic alkalosis. Bartter's syndrome, in contrast to Gitelman's syndrome, is associated with *hyper*calciuria. Bartter's syndrome does not generally occur in adults.

Renal Failure

Although renal failure is a well-known cause of metabolic *acidosis,* renal failure may *maintain* a metabolic alkalosis because the failing kidney is unable to excrete bicarbonate. In order for the kidney to correct a metabolic alkalosis by excreting excess bicarbonate, there must be an adequate glomerular filtration rate. Obviously, if one cannot filter bicarbonate, then it cannot be excreted!.

Respiratory Compensation for Metabolic Alkalosis

The hydrogen ion concentration of the ECF is determined by the ratio of the P_{CO_2}, which is controlled by the lungs, to the $[HCO_3^-]$, which is controlled by the kidneys, according to the relation:

$$[H^+] \propto P_{CO_2}/[HCO_3^-]$$

A metabolic alkalosis is a process which causes a primary increase in $[HCO_3^-]$. The respiratory compensation for a metabolic alkalosis is decreased ventilation, which produces a secondary increase in P_{CO_2}. This returns the $P_{CO_2}/[HCO_3^-]$ ratio (and therefore the hydrogen ion concentration) toward the normal range. The lungs do not bring the hydrogen ion concentration into the normal range, but only *toward* the normal range. By how much should the P_{CO_2} increase in compensation for a metabolic alkalosis? The quantitative answer is obtained by using the formula for expected respiratory compensation for a metabolic alkalosis. That is: the P_{CO_2} should be equal to

$$P_{CO_2} = 40 + .7 \times ([HCO_3^-{}_{(measured)}] - [HCO_3^-{}_{(normal)}])$$

where $[HCO_3^-{}_{(measured)}]$ is the patient's measured bicarbonate and $[HCO_3^-{}_{(normal)}]$ is the normal bicarbonate concentration, roughly 24–26 mEq/L.

What if the measured P_{CO_2} differs from this value? A significant deviation means that there is also a respiratory disorder present, because the P_{CO_2} is not behaving as we would expect. If the measured P_{CO_2} is higher than predicted by the formula, then a coexisting respiratory acidosis is present. If the measured P_{CO_2} is lower than predicted, then a coexisting respiratory alkalosis is present. Note that this formula is an approximation, and there is an especially wide patient to patient variation in the respiratory response to metabolic alkalosis. Therefore, for metabolic alkalosis, we should allow the P_{CO_2} to be ±5 mm Hg off from that predicted by the formula. A significant deviation in either direction from the value predicted by the expected compensation formula, however, indicates that in addition to a metabolic alkalosis, there is also a respiratory disorder present.

Treatment of Metabolic Alkalosis

The treatment of metabolic alkalosis (Fig. 8-2) should be directed at

- The underlying cause of the metabolic alkalosis
- The cause of the renal retention of HCO_3^-

FIGURE 8-2. Treatment of Metabolic Alkalosis

Cause	Treatment	Remarks
Vomiting/nasogastric suction	ECFV replacement with NaCl (.9% normal saline).	K^+ deficit and sometimes Mg^{++} deficit also present.
Diuretic therapy	ECFV replacement with NaCl (.9% normal saline).	K^+ deficit and sometimes Mg^{++} deficit also present.
Posthypercapneic	ECFV replacement with NaCl (.9% normal saline).	Correct K^+ deficit. Patients may be receiving diuretics or steroids.
Chronic diarrhea	Careful ECFV replacement with NaCl if significantly volume depleted (.9% normal saline).	K^+ deficits and sometimes Mg^{++} deficits also present. Diagnose the cause of the diarrhea.
Potassium depletion (maintaining alkalosis)	Potassium chloride replacement.	Potassium deficits are often quite significant.
Mineralocorticoid excess	Potassium chloride replacement.	Identify and treat the specific underlying disease state.
Bartter's syndrome	Potassium chloride replacement.	Nonsteroidal antiinflammatory drugs, amiloride as adjunctive agents
Alkalosis associated with ECFV overload/renal failure	NaCl contraindicated. Potassium infusion potentially dangerous.	HCl Dialysis

Treatment of ECFV Depletion

In saline-responsive metabolic alkalosis, volume expansion with 0.9% saline should correct the alkalosis. Remember that giving hypotonic fluids to patients with ECFV depletion can lead to dangerous hyponatremia. Coexisting potassium deficits are almost always present and should be corrected as well. As the ECFV deficit is replaced, an alkaline diuresis occurs, with reduction of the serum HCO_3^- concentration. This may further worsen the existing potassium deficit, because excess HCO_3^- "carries" sodium with it to the collecting tubule, causing increased sodium-potassium exchange.

In posthypercapnic alkalosis, appropriate management consists of managing ventilatory status and by correction of ECFV depletion which permits renal excretion of HCO_3^-.

Treatment of Severe Potassium Depletion

Profound potassium depletion to <2.0 mEq/L, regardless of cause, can act to maintain metabolic alkalosis. Replacement of the potassium deficits allows correction of the metabolic alkalosis.

Treatment of Mineralocorticoid Excess

If primary mineralocorticoid excess is present, the underlying disorder should be sought. NaCl replacement is generally contraindicated because these patients are generally already volume expanded and hypertensive. Patients require potassium chloride replacement in addition to specific therapy aimed at the underlying disease entity.

Treatment of Bartter's Syndrome

Potassium chloride supplements alone are usually not sufficient to match the ongoing potassium losses in Bartter's syndrome. Nonsteroidal antiinflammatory drugs (which counteract kaliuresis) and the potassium-sparing diuretic amiloride have been used as adjunctive measures with some success. With treatment, the serum potassium may stabilize at a level greater than 3.0 mEq/L.

Treatment of Metabolic Alkalosis in Volume Overload and Renal Failure

The treatment of metabolic alkalosis in volume overloaded patients and those with renal failure is complex. Saline administration is generally contraindicated in this setting. In the patient with renal failure, dialysis against a low bicarbonate dialysate may help reverse the alkalosis. If a patient's alkalosis is acutely life-threatening, dilute HCl can be used under extreme circumstances in an intensive care setting to reduce the patient's serum $[HCO_3^-]$ promptly.

Exercises

1. A 40-year-old man is admitted with the following chemistries: sodium 140 mEq/L, chloride 86 mEq/L, bicarbonate 40 mEq/L, potassium 3.0 mEq/L, glucose 120 mg/dl, BUN 32 mg/dl, Cr 1.4 mg/dl. Arterial blood gas: pH 7.52, P_{CO2} 51 mm Hg, HCO_3^- 40 mEq/L. What is your general approach to this patient?
   Answer: The diagnosis of acid-base disorders requires a systematic approach to identify *all* the disorders present in a given patient. Chapter 9 describes a simple, three-step method to use for every single acid-base problem. Do not worry if you do not understand all three steps right now. Just follow along.

Step 1: Identify a single disorder. The pH is high and the bicarbonate is also high.
Therefore, metabolic alkalosis is present.

Step 2: See if the compensation is correct. If the compensatory change in P_{CO_2} in response to the metabolic alkalosis is not as predicted by the formula, then a respiratory disorder is present. For a metabolic alkalosis, the P_{CO_2} should be

$$P_{CO_2} = 40 + .7 \times ([HCO_3^-_{(measured)}] - [HCO_3^-_{(normal)}])$$
$$= 40 + .7 \ (40\text{--}24) = 51.2 \text{ mm Hg.}$$

This assumes a normal $[HCO_3^-]$ of 24 mEq/L. The measured P_{CO_2} of 51 mm Hg is equal to that predicted by the formula. Therefore, compensation is appropriate, and no respiratory disorder is present. Remember to use values for both the P_{CO_2} and the $[HCO_3^-]$ from the arterial blood gas for Step 2.

Step 3: Calculate the anion gap using serum values:

$$AG = [Na^+] - ([Cl^-] + [HCO_3^-])$$
$$= 140 - (86 + 40) = 14$$

The anion gap is normal. We are finished.

This analysis tells us that the only acid-base disorder present is a metabolic alkalosis. Now, we must turn our attention to finding and correcting the cause for the abnormal renal retention of bicarbonate. We perform a careful history and physical, which includes checking for ECFV depletion due to vomiting, nasogastric suction, chronic diarrhea/laxative use, and diuretics, as well as signs of secondary hyperaldosteronism (congestive heart failure and cirrhosis). Hypertension suggests renovascular disease, primary hyperaldosteronism, or Cushing's syndrome, although primary hypertension may occur without any of these disorders. Remember that profound potassium depletion alone may cause renal retention of bicarbonate.

The urine chloride is helpful in diagnosing metabolic alkalosis caused by ECFV depletion. It is generally low (<10 mEq/L) in cases of metabolic alkalosis maintained by ECFV depletion. The urine sodium, which is also generally low in cases of volume depletion, is less helpful in cases of metabolic alkalosis caused by ECFV depletion. The reason is that excess bicarbonate that is not reabsorbed by the proximal tubule acts as a non-reabsorbable anion, bringing sodium to the collecting tubule. This results in more sodium in the urine, and consequently, a higher urine sodium. *Therefore, the urine chloride is a better test than the urine sodium for extracellular volume depletion when metabolic alkalosis is present.*

2. A 20-year-old woman is admitted with the following serum chemistries: sodium 140 mEq/L, chloride 90 mEq/L, bicarbonate 34 mEq/L, potassium 3.0 mEq/L, glucose 120 mg/dl, BUN 30 mg/dl. Arterial blood gas: pH 7.48, P_{CO_2} 47 mm Hg, HCO_3^- 34 mEq/L. She tells you that she thinks she

might have "Gitelman's syndrome." She consistently denies vomiting, laxatives, or diuretics. What do you do?

Answer: First, see what disorders you have.

Step 1: Identify the most apparent disorder. The pH is high and the bicarbonate is also high. Therefore, metabolic alkalosis is present.

Step 2: See if the compensation is correct (if the compensation of the P_{CO2} for the metabolic alkalosis is not what is predicted from the formula, then a respiratory disorder is present). For a metabolic alkalosis, the P_{CO2} should be

$$P_{CO2} = 40 + .7 \times ([HCO_3^-{}_{(measured)}] - [HCO_3^-{}_{(normal)}])$$
$$= 40 + .7 \times (34-24) = 47 \text{ mm Hg.}$$

This assumes a normal $[HCO_3^-]$ of 24 mEq/L. The measured P_{CO2} of 47 mm Hg is equal to that predicted by the formula. Therefore, compensation is appropriate, so no respiratory disorder is present.

Step 3: Calculate the anion gap:

$$AG = [Na^+] - ([Cl^-] + [HCO_3^-])$$
$$= 140 - (90 + 34) = 16$$

The AG is normal. Metabolic alkalosis is the only disorder present.

We begin our clinical evaluation. There is a drop in blood pressure and increase in heart rate with standing, indicating ECFV depletion. The presence of ECFV depletion suggests vomiting, diuretics, chronic diarrhea and laxative abuse. The spot urine sodium is 40 mEq/L, and the chloride 5 mEq/L. The urine potassium is 10 mEq/ gm creatinine. Now what?

Answer: These studies suggest that ECFV depletion is present, either from surreptitious vomiting, diuretic abuse, or from chronic laxative abuse. The urine concentrations of chloride and potassium are low, consistent with renal conservation of these ions. The urine sodium is higher than we might normally expect for ECFV depletion, but during metabolic alkalosis HCO_3^- "carries" sodium with it, elevating the urine sodium concentration. The urine chloride is a more reliable indicator of ECFV depletion in states of metabolic alkalosis. Several repeated urine tests for diuretics and a stool sample for laxatives could be helpful. Gitelman's syndrome is associated with normotension and *high* urine chloride, sodium, and potassium concentrations because these ions are being lost in the urine. Gitelman's syndrome is also rare, so always suspect the more common causes of metabolic alkalosis first.

3. Same patient as in exercise 2. The screens are negative, but a follow-up spot urine shows urine sodium 65 mEq/L, chloride 75 mEq/L, and urine potassium 40 mEq/gram creatinine. What do you do?

   Answer: This pattern suggests diuretic abuse. Between doses of diuretics, the urine chloride and potassium concentrations are low, consistent with renal conservation of these ions. But the urine concentrations of chloride and potassium are high when the dose of diuretic is acting. The key here would be to *repeat* the urine screen for diuretics at a time when the urine values suggest a diuretic effect.

4. A 50-year-old woman is admitted with the following serum chemistries: sodium 140 mEq/L, chloride 90 mEq/L, bicarbonate 34 mEq/L, potassium 3.0 mEq/L, glucose 120 mg/dl, BUN 20 mg/dl, creatinine 1.1 mg/dl. Arterial blood gas: pH 7.48, P_{CO2} 47 mm Hg, HCO_3^- 34 mEq/L. The urine sodium is 65 mEq/L, urine chloride 70 mEq/L, and urine potassium 40 mEq/gm creatinine. The blood pressure is 196/124 mm Hg. She is not obese, and there are no stigmata of Cushing's syndrome, liver disease, or congestive heart failure. She is not taking diuretics or laxatives. What is your differential diagnosis?

Answer: The differential diagnosis includes the causes of metabolic alkalosis listed under mineralocorticoid excess syndromes. Possibilities in this very hypertensive patient include primary hyperaldosteronism, renovascular hypertension, and malignant hypertension (which may result from either primary hypertension or a number of causes of secondary hypertension).

Therapeutic or surreptitious use of thiazide or loop diuretics could also give you the same urine and serum electrolyte picture in a patient with primary hypertension and should be considered. Gitelman's syndrome is associated with *normotension* and high urine concentrations of chloride, sodium, and potassium because these ions are being lost in the urine. Hypertension of this degree essentially excludes Gitelman's syndrome.

5. A 60-year-old man is admitted with protracted vomiting. His chemistries: sodium 140 mEq/L, chloride 84 mEq/L, bicarbonate 40 mEq/L, potassium 3.6 mEq/L, glucose 120 mg/dl, BUN 80 mg/dl, creatinine 9.8 mg/dl. Arterial blood gas: pH 7.52, P_{CO2} 51 mm Hg, HCO_3^- 40 mEq/L. The patient was treated with 3 L 0.9% saline while the laboratory studies were being done. The patient now has bibasilar rales and states that he is short of breath. There is no urine output. What do you do?

Answer: This patient has metabolic alkalosis associated with severe renal failure. The renal failure will not allow correction of the alkalosis by appropriate bicarbonate excretion. This is a complex situation and is best handled with the help of a nephrologist, who can dialyze the patient acutely if necessary. Saline administration is contraindicated in this setting. In the patient with renal failure, dialysis against a low bicarbonate dialysate may reverse the alkalosis. Rarely, dilute HCl may be used under extreme circumstances in an intensive care unit setting.

CHAPTER 9. MIXED ACID-BASE DISORDERS

Just as it is important to have a consistent, systematic way to view a chest X-ray or interpret an ECG, it is important to have a systematic approach to analyzing acid-base chemistries. This chapter describes a simple, step-by-step method, which can be used every time acid-base chemistries are obtained.

Mixed (complex) acid-base disorders are cases in which two or more *independent* acid-base disorders occur simultaneously. A step-by-step approach is important because two or more disorders may sometimes "mask" each other. For example, a metabolic acidosis and metabolic alkalosis may offset each other, leading to a relatively normal pH and $[HCO_3^-]$. Alternatively, a severe predominant disorder may overshadow a milder disorder. As we shall see, dangerous acid-base disorders can be overlooked altogether unless there is a systematic, step-by-step approach to examining acid-base chemistries.

Diagnosis of Acid-Base Disorders: A Three-Step Approach

A three-step approach to acid-base disorders (see **Fig. 9-1**) will be described in the remainder of this chapter:

Step 1: Identify the most obvious disorder (metabolic acidosis, metabolic alkalosis, respiratory acidosis, or respiratory alkalosis) by looking at the pH, P_{CO2} and $[HCO_3^-]$. If more than one acid-base disorder is apparent right away, just pick the "worst" disorder to start with. What if they all look severe? Then just pick one. A nice feature of this approach is that you can start anywhere, and the solution to the case will always work out the same.

Step 2: Apply the formula for the expected compensation for the disorder you have identified to determine if a second disorder is present. Once you identify a disorder, the general question is: Is the compensation for this disorder appropriate?

• In metabolic disorders, the $[HCO_3^-]$ is abnormal and we want to see if there is also a coexisting respiratory disorder. We ask: What should the P_{CO2} be after compensation? If the P_{CO2} differs significantly from that predicted by the formula for compensation, then a coexisting respiratory disorder is present.

FIGURE 9-1. Three Step Approach to Acid-Base Disorders

Step 1: Identify *the most apparent* disorder.

Disorder	pH	P_{CO2}	HCO_3^-
Metabolic acidosis	Decreased	Decreased (secondary)	Decreased (primary)
Metabolic alkalosis	Increased	Increased (secondary)	Increased (primary)
Respiratory acidosis	Decreased	Increased (primary)	Increased (secondary)
Respiratory alkalosis	Increased	Decreased (primary)	Decreased (secondary)

Step 2: Apply the formulas to determine if compensation is appropriate. If not, a second disorder co-exists.

Metabolic acidosis: $P_{CO2} = 1.5 \times [HCO_3^-] + 8$

Metabolic alkalosis: $P_{CO2} = 40 + .7 \times ([HCO_3^-_{(measured)}] - [HCO_3^-_{(normal)}])$

Respiratory acidosis:

Acute: $[HCO_3^-]$ increases by 1 mEq/L for every 10 mm Hg increase in P_{CO2}

Chronic: $[HCO_3^-]$ increases by 3.5 mEq/L for every 10 mm Hg increase in P_{CO2}

Respiratory alkalosis:

Acute: $[HCO_3^-]$ decreases by 2 mEq/L for every 10 mm Hg decrease in P_{CO2}

Chronic: $[HCO_3^-]$ decreases by 5 mEq/L for every 10 mm Hg decrease in P_{CO2}

Step 3: Calculate the anion gap.

- $AG = [Na^+] - ([Cl^-] + [HCO_3^-])$
- The normal AG is 9-16 mEq/L.
- If AG > 20 mEq/L, high AG acidosis is probably present.
- If AG > 30 mEq/L, high AG acidosis is almost certainly present.
- For lactic acidosis, the ratio of the increase in the anion gap to the decrease in the HCO_3^- averages approximately 1.5.
- In ketoacidosis, the ratio of the increase in the anion gap to the decrease in the HCO_3^- averages approximately 1.0.

- In respiratory disorders, the P_{CO2} is abnormal and we want to see if there is also a coexisting metabolic disorder. We ask: What should the $[HCO_3^-]$ be after compensation? If the $[HCO_3^-]$ differs significantly from that predicted by the formula for compensation, then a coexisting metabolic disorder is present.

Apply the formula for the disorder you have identified to see if the compensation is correct. If the compensation is *not* what is predicted by the formula, then an additional disorder is present.

Step 3: Calculate the anion gap. The normal value of the anion gap used in this book is 9–16 mEq/L, although many hospitals may prefer to use the smaller range 10–14. If the calculated anion gap is normal, you are finished. The presence of an increased anion gap is a powerful clue to the diagnosis of metabolic acidosis. If the anion gap is increased above 20 mEq/L, then an anion gap metabolic acidosis is probably present. If the anion gap is increased above 30 mEq/L, then an anion gap metabolic acidosis is almost certainly present, regardless of the pH and $[HCO_3^-]$.

If a high anion gap acidosis due to lactic acidosis or ketoacidosis is present, then it may be helpful to compare the change in the anion gap to the change in the bicarbonate concentration. By doing this, one may identify an additional "hidden" metabolic disorder, either a metabolic alkalosis or a normal anion gap metabolic acidosis.

Step 1: Identify One Disorder.

Look at the pH, P_{CO_2} and $[HCO_3^-]$ to identify the most apparent acid-base disorder. In general:

- If the pH is low (<7.35), either a metabolic acidosis or a respiratory acidosis is present; if the $[HCO_3^-]$ is low: metabolic acidosis; if the P_{CO_2} is high: respiratory acidosis.
- If the pH is high (>7.45), either a metabolic alkalosis or a respiratory alkalosis is present; if the $[HCO_3^-]$ is high: metabolic alkalosis; if the P_{CO_2} is low: respiratory alkalosis.
- If the pH is normal, but either the $[HCO_3^-]$ or the P_{CO_2}, (or both) is abnormal, then pick the most abnormal of the $[HCO_3^-]$ or P_{CO_2}. For example: pH 7.40, P_{CO_2} 60 mm Hg, HCO_3 36 mEq/L. Both the P_{CO_2} and the $[HCO_3^-]$ are abnormal. Because the pH is normal in this case, you could start by diagnosing *either* a metabolic alkalosis ($[HCO_3^-]$ 36 mEq/L) *or* a respiratory acidosis (P_{CO_2} 60 mm Hg). This method will allow you to start either way.

Step 2: Apply the Formulas to See If Compensation Is Correct.

Apply the formulas for expected compensation to determine if a second disorder is present. This section deals with what the formulas for expected compensation to simple disorders mean and how to use them. Once you identify a disorder, the general question is: Is the compensation close to that predicted by the formula for expected compensation? Once you have made a diagnosis of one disorder, then apply the formula for that specific disorder to see if the compensation is appropriate. For metabolic disorders ask: What should the P_{CO_2} be after compensation? For respiratory disorders ask: What should the $[HCO_3^-]$ be after compensation? The formulas give approximations for the expected compensation for acid-base disorders. If the compensation is not consistent with the given formula, then a *second* disorder is present.

Remember to use the values of both the P_{CO_2} and the $[HCO_3^-]$ from the arterial blood gas (ABG) for purposes of determining if compensation is appropriate (Step 2). Also, remember to use serum values to calculate the anion gap (Step 3). In this book, the serum bicarbonate and the calculated bicarbonate from the ABG are almost always equal, but this is not always the case in clinical practice.

Metabolic Acidosis

The hydrogen ion concentration of ECF is determined by the ratio of the P_{CO_2} (which is controlled by the lungs) to the $[HCO_3^-]$ (which is controlled by the kidneys) according to the relation:

$$[H^+] \propto P_{CO_2}/[HCO_3^-]$$

A metabolic acidosis is a process that causes a primary decrease in $[HCO_3^-]$. The respiratory compensation for a metabolic acidosis is increased ventilation, which produces a secondary decrease in P_{CO_2}. This returns the $P_{CO_2}/[HCO_3^-]$ ratio (and therefore the hydrogen ion concentration) toward the normal range. Typically, the lungs do not return the hydrogen ion concentration all the way into the normal range, but only toward the normal range. What should the P_{CO_2} be after compensation for a metabolic acidosis? The quantitative answer to this question is obtained by using the formula for expected respiratory compensation for a metabolic acidosis. That is, the P_{CO_2} should be equal to:

$$P_{CO_2} = 1.5 \times [HCO_3^-] + 8$$

What if the measured P_{CO_2} differs from this value? A significant difference means that there is also a respiratory disorder in addition to the metabolic acidosis, because the P_{CO_2} is not behaving as we would expect. If the measured P_{CO_2} is higher than predicted by the formula, there is a coexisting respiratory acidosis. If the measured P_{CO_2} is lower than predicted, there is a coexisting respiratory alkalosis. This formula is approximate, and we should allow the measured P_{CO_2} to be ± 2 mm Hg off from that predicted by the formula. A more significant deviation in either direction from the value predicted by the formula, however, indicates that in addition to a metabolic acidosis, there is also a respiratory disorder present.

Metabolic Alkalosis

The hydrogen ion concentration of the ECF, as mentioned, is determined by the ratio of the P_{CO_2}, which is controlled by the lungs, to the $[HCO_3^-]$, which is controlled by the kidneys, according to the relation:

$$[H^+] \propto P_{CO_2}/[HCO_3^-]$$

A metabolic alkalosis is a process which causes a primary increase in $[HCO_3^-]$. The respiratory compensation for a metabolic alkalosis is decreased ventilation, which produces a secondary increase in P_{CO_2}. This returns the $P_{CO_2}/[HCO_3^-]$ ratio (and therefore the hydrogen ion concentration) toward the normal range. The lungs do not bring the hydrogen ion concentration into the normal range, but only *toward* the normal range. By how much should the P_{CO_2} increase in compensation for a metabolic alkalosis? The quantitative answer to this is obtained by using the formula for expected respiratory compensation for a metabolic alkalosis. That is: the P_{CO2} should be equal to

$$P_{CO2} = 40 + 0.7 \times ([HCO_3^-_{(measured)}] - [HCO_3^-_{(normal)}])$$

where $[HCO_3^-_{(measured)}]$ is the patient's measured bicarbonate and $[HCO_3^-_{(normal)}]$ is the normal bicarbonate concentration, roughly 24–26 mEq/L.

What if the measured P_{CO2} differs from this value? A significant deviation means that there is also a respiratory disorder present, because the P_{CO2} is not behaving as we would expect. If the measured P_{CO2} is higher than predicted by the formula, then a coexisting respiratory acidosis is present. If the measured P_{CO2} is lower than predicted, then a coexisting respiratory alkalosis is present. This formula is an approximation, and there is an especially wide patient to patient variation in the respiratory response to metabolic alkalosis. Therefore, for metabolic alkalosis, we should allow the P_{CO2} to be ± 5 mm Hg off from that predicted by the formula. A more significant deviation in either direction from the value predicted by the expected compensation formula, however, indicates that in addition to a metabolic alkalosis, there is also a respiratory disorder present.

The maximum value P_{CO2} can reach in compensating for a metabolic alkalosis is about 55 mm Hg. A P_{CO2} of more than 55 mm Hg generally implies that a respiratory acidosis is also present, regardless of the $[HCO_3^-]$.

Respiratory Disorders

A respiratory acidosis is a process that causes a primary increase in the P_{CO2}. A respiratory alkalosis is a process that causes a primary decrease in the P_{CO2}. Respiratory disorders are divided into acute and chronic. Acute means minutes to an hour or so and chronic means more than 24–48 hours. The reason for this seemingly arbitrary distinction is that the full renal compensation for respiratory disorders takes at least 24–48 hours. That is, when a metabolic acidosis develops, the increase in minute ventilation comprising the respiratory compensation occurs rapidly and the P_{CO2} falls right away. In a respiratory disorder, however, the kidneys take at least 24–48 hours to fully adjust the $[HCO_3^-]$ to its new compensatory value. A short way of saying this is that the lungs adjust the P_{CO2} much faster than the kidneys can adjust the $[HCO_3^-]$!

For example, in the case of a respiratory acidosis, the primary derangement is an increase in P_{CO2}. The renal compensation for a respiratory acidosis leads to a secondary increase in $[HCO_3^-]$. Although the renal compensation tending to increase the $[HCO_3^-]$ begins right away, it takes at least 24–48 hours for the kidney to increase the $[HCO_3^-]$ to its new steady state value. Therefore, the formulas for predicted compensation for respiratory disorders depend upon whether the disorder is acute (minutes to an hour or so) or chronic (24–48 or more hours, at which time the renal compensation has fully taken place).

For an acute respiratory acidosis, the serum HCO_3^- concentration rises about 1 mEq/L for every 10 mm Hg increase in P_{CO2}. For a chronic respiratory acidosis (after 24–48 hours), the serum HCO_3^- bicarbonate concentration rises about 3.5 mEq/L for every 10 mm Hg increase of P_{CO2}. For respiratory alkalosis the P_{CO2} falls, and therefore the compensation is for the serum HCO_3^- concentration to decrease as well. How much? For acute respiratory alkalosis (minutes to an hour or so), the serum HCO_3^- concentration will fall about 2 mEq/L for every 10 mm Hg drop in P_{CO2}. For chronic respiratory alkalosis (more than 24–48 hours), the serum HCO_3^-concentration will fall by about 5 mEq/L for every 10 mm Hg fall in P_{CO2}.

In summary, for respiratory disorders:

Respiratory acidosis:
  Acute: $[HCO_3^-]$ increases by 1 mEq/L for every 10 mm Hg increase in P_{CO2}.
  Chronic: $[HCO_3^-]$ increases by 3.5 mEq/L for every 10 mm Hg increase in P_{CO2}.
Respiratory alkalosis:
  Acute: $[HCO_3^-]$ decreases by 2 mEq/L for every 10 mm Hg decrease in P_{CO2}.
  Chronic: $[HCO_3^-]$ decreases by 5 mEq/L for every 10 mm Hg decrease in P_{CO2}.

Example 1: A patient's P_{CO2} increases from 40 mm Hg to 60 mm Hg during a chronic respiratory acidosis. For a chronic respiratory acidosis, the serum $[HCO_3^-]$ rises about 3.5 mEq/L for every 10 mm Hg increase of P_{CO2}. The compensatory change in serum $[HCO_3^-]$ would be an increase of $2 \times 3.5 = 7$ mEq/L. The multiplication factor **2** is used because the increase in P_{CO2} is 20 ($= 2 \times 10$ increase in P_{CO2}).

Example 2: A patient has a P_{CO2} of 40 and a $[HCO_3^-]$ of 24 mEq/L. An acute respiratory alkalosis develops and the P_{CO2} falls to 20 mm Hg. What should the $[HCO_3^-]$ be after compensation?
Answer: The $[HCO_3^-]$ should decrease by $2 \times 2 = 4$ mEq/L. The multiplication factor **2** is used because a decrease in P_{CO2} of 20 mm Hg $= 2 \times 10$ decrease in P_{CO2}. The $[HCO_3^-]$ should be $24 - 4 = 20$ mEq/L after compensation for the acute respiratory alkalosis.

Example 3: A patient has a P_{CO2} of 40 and a $[HCO_3^-]$ of 24 mEq/L. An acute respiratory acidosis develops and the P_{CO2} rises to 70 mm Hg. What should the $[HCO_3^-]$ be after compensation?
Answer: The $[HCO_3^-]$ should increase by $3 \times 1 = 3$ mEq/L. The multiplication factor **3** is used because an increase in P_{CO2} of 30 mm Hg $= 3 \times 10$ increase in P_{CO2}. The $[HCO_3^-]$ should be $24 + 3 = 27$ mEq/L after compensation for the acute respiratory acidosis.

Example 4: A patient has a P_{CO2} of 40 and a $[HCO_3^-]$ of 24 mEq/L. A chronic respiratory acidosis develops and the P_{CO2} rises to 70 mm Hg. What should the $[HCO_3^-]$ be after compensation ($>24-48$ hours)?
Answer: The $[HCO_3^-]$ should increase by $3 \times 3.5 = 10.5$ mEq/L. The multiplication factor **3** is used because an increase in P_{CO2} of 30 mm Hg $= 3 \times 10$ increase in P_{CO2}. The $[HCO_3^-]$ should be $24 + 10.5 = 34.5$ mEq/L after compensation for the chronic respiratory acidosis.

Step 3: Calculate the Anion Gap.

$$AG = [Na^+] - ([Cl^-] + [HCO_3^-])$$

The normal value of the anion gap used in this book is 9–16 mEq/L. If the anion gap is normal, then you are finished. Count up the acid-base disorders that you have identified and take a bow. The presence of an increased anion gap is a powerful clue to the diagnosis of metabolic acidosis. Here are some general guidelines on using the AG to diagnose high anion gap metabolic acidosis:

- If the anion gap is greater than or equal to 30 mEq/L, then there is a high anion gap acidosis, regardless of the $[HCO_3^-]$ and pH.
- If the anion gap is greater than 20 mEq/L, then there *probably* is a high anion gap metabolic acidosis, regardless of the $[HCO_3^-]$ and pH.
- Anion gaps in the range 16-20 mEq/L are abnormal, but may be due to other things besides an anion gap metabolic acidosis.
- If the anion gap is normal, then you are finished with the last step.

Comparing the Change in the Anion Gap to the Change in Bicarbonate

If you can perform the three steps that have been described in the preceding text, you are in very good shape for solving most of the mixed acid-base problems that you will encounter in clinical practice. There is one additional step that applies only to cases in which there is a high anion gap acidosis due to lactic acidosis or ketoacidosis. This step is sometimes useful to detect additional "hidden" metabolic disorders.

If you make the diagnosis of a high anion gap metabolic acidosis due to lactic acidosis or ketoacidosis in Step 3, it may be helpful to compare the change in the AG to the change in the $[HCO_3^-]$. One might suppose that in a high anion gap metabolic acidosis, there would be a correlation between the increase in the anion gap, which is caused by the addition of the anion to the ECF, and the decrease in the bicarbonate, which is caused by the titration of HCO_3^- by the hydrogen ion. According to the equation

$$AG = [Na^+] - ([Cl^-] + [HCO_3^-])$$

131

one could logically expect that if the AG increases because of a high anion gap acidosis, the HCO_3^- concentration would decrease by an equal amount. For example, if a lactic acidosis or diabetic ketoacidosis increases the anion gap by 15 mEq/L, the HCO_3^- concentration might be expected to fall by an equal amount, 15 mEq/L.

A one-to-one relationship between the increase in the anion gap and the decrease in bicarbonate is often *not* the case, however. One reason is that hydrogen ion is buffered intracellularly and by bone as well as by the HCO_3^- in extracellular fluid. Simply put: HCO_3^- does not have to buffer all the hydrogen ion by itself, but "gets help" from other buffer systems. Therefore, the $[HCO_3^-]$ may decrease by an amount *less* than the increase in the anion gap. For lactic acidosis, the ratio of the increase in the AG to the decrease in the $[HCO_3^-]$ is not usually 1.0, but on the average may actually be closer to 1.5 because of this extra buffering of hydrogen ion outside the ECF. That is, for lactic acidosis, approximately:

$$\text{Change in AG/Change in } [HCO_3^-] = 1.5$$

or, rearranging:

$$\text{Change in } [HCO_3^-] = \text{Change in AG}/1.5$$

Using this very rough formulation, we might expect that if a lactic acidosis increases the AG by 15 mEq/L, then the $[HCO_3^-]$ would fall by about: Change in AG/1.5 = 15/1.5 = 10 mEq/L, not 15 mEq/L.

For ketoacidosis, the ratio of the increase in the AG to the decrease in the $[HCO_3^-]$ *is* closer to 1.0, perhaps because some ketoanions, which constitute the increase in the AG, may be lost in the urine. Therefore, for ketoacidosis, approximately:

$$\text{Change in } [HCO_3^-] = \text{Change in AG}$$

It should be carefully restated that this is a *very rough* way to estimate the expected fall in $[HCO_3^-]$ for a given increase in AG when there is a lactic acidosis or a ketoacidosis. For uremic acidosis and the other causes of high anion gap metabolic acidosis, the relationship between the increase in the AG and the decrease in the bicarbonate is unpredictable.

How can we use this information in the setting of lactic acidosis or ketoacidosis? A measured $[HCO_3^-]$ much higher than predicted by the increase in anion gap is a clue that a "hidden" metabolic alkalosis may also be present. A measured $[HCO_3^-]$ much less than predicted by the increase in anion gap is a clue that a "hidden" normal anion gap metabolic acidosis may also be present.

When I diagnose a high anion gap acidosis due to a lactic acidosis or a ketoacidosis, I compare the predicted fall in bicarbonate (based upon what you would expect from the increase in anion gap) to the actual fall in bicarbonate, then use the following guidelines:

- A measured $[HCO_3^-]$ much higher than predicted by the increase in anion gap is a clue that a "hidden" metabolic alkalosis may also be present.
- A measured $[HCO_3^-]$ much less than predicted by the increase in anion gap is a clue that a "hidden" normal anion gap metabolic acidosis may also be present.

Example 1: A patient begins with a serum $[HCO_3^-]$ of 24 and an AG of 12. She develops a *lactic* acidosis. The AG increases from 12 to 22. Approximately, what would you expect the $[HCO_3^-]$ to be?
Answer: The 10 mEq/L of H^+ is buffered not only by extracellular HCO_3^-, but also by intracellular buffers and by bone. The $[HCO_3^-]$ would be expected to decreases by about: Change in AG/1.5 = 10/1.5 = 6.7 mEq/L. The expected $[HCO_3^-]$ is therefore: 24 − 6.7 = 17.3 mEq/L.

The increase of 10 mEq/L in the AG is accompanied by an opposing decrease in the $[HCO_3^-]$ of 6.7 mEq/L. I realize that the decimal places may look strange with regard to the HCO_3^- concentration, especially because this is only an approximate method anyway. I am going to leave the decimals in, however, so you can follow the calculations as we go. I know that it really doesn't make much sense to talk about a $[HCO_3^-]$ of 17.3 mEq/L.

Example 2: A patient begins with a serum $[HCO_3^-]$ of 24 mEq/L and an anion gap of 12 mEq/L. A ketoacidosis develops, and the anion gap increases from 12 to 22 mEq/L. What should the resulting $[HCO_3^-]$ be?
Answer: The AG has increased by 10 mEq/L. Therefore, the $[HCO_3^-]$ should decrease by about 10 mEq/L (remember that the Change in AG/Change in $[HCO_3^-]$ averages about 1.0 in ketoacidosis). Therefore, the bicarbonate would be expected to fall from 24 mEq /L to 14 mEq/L (24 − 10 = 14 mEq/L). The increase in AG should be associated with an opposing change in the $[HCO_3^-]$.

Example 3: A patient starts with AG 12, serum $[HCO_3^-]$ 24. ABG: pH 7.40, $[HCO_3^-]$ 24 and P_{CO2} 40. A lactic acidosis develops, and the AG rises from 12 to 32 mEq/L. The $[HCO_3^-]$ does not fall: It stays at 24 mEq/L. The pH remains 7.40 and the P_{CO2} is 40. What has happened?
Answer: Beginning with the 3-step approach:
Step 1: Everything looks normal. No acid-base disorder so far.
Step 2: The P_{CO2} is normal and appropriate for the normal $[HCO_3^-]$ of 24. Therefore, no respiratory disorder is present.
Step 3: The AG has increased by 20 mEq/L. By considering the change in the anion gap, 20 mEq/L, the predicted $[HCO_3^-]$ would fall by about: Change in AG/1.5 = 20/1.5 = 13.3 mEq/L. The expected $[HCO_3^-]$ would be 24 − 13.3 = 10.7 mEq/L. But we can see that

the patient's $[HCO_3^-]$ has not fallen but has remained at 24 mEq/L. Why? There must be something which is "pushing the $[HCO_3^-]$ up" by 13.3 mEq/L. What is it?

Answer: It is a metabolic alkalosis. The severe anion gap metabolic acidosis is "masked" by a metabolic alkalosis of equal severity. If we did not calculate the anion gap and then compare the predicted fall in bicarbonate (based upon what you would expect given the increase in anion gap) to the actual fall in bicarbonate, we would have missed these *two* independent and serious disorders. I forgot to tell you that this patient had been vomiting for the last two days along with developing lactic acidosis.

Example 4: A patient starts with AG 12, serum $[HCO_3^-]$ 24. ABG: pH 7.40, $[HCO_3^-]$ 24, and P_{CO2} 40. A lactic acidosis develops, and the new values are AG 28, $[HCO_3^-]$ 22, P_{CO2} 39, and pH 7.37. What disorders are present?
Answer:

Step 1: The pH and the $[HCO_3^-]$ have fallen: metabolic acidosis. This appears very mild at first glance.

Step 2: What should the P_{CO2} be after compensation? Use the formula in **Fig. 9-1.** The P_{CO2} should be $1.5 \times 22 + 8 = 41$. The patient's P_{CO2} is 39. This is well within the predicted range for P_{CO2}. Therefore, there is no respiratory disorder present.

Step 3: The change in the anion gap is an increase of 16 mEq/L. If there were only a high anion gap acidosis present, then we would expect the $[HCO_3^-]$ to decrease roughly by an amount: Change in AG/1.5 = $16/1.5 = 10.7$ mEq/L. The expected $[HCO_3^-]$ would be decreased to $24 - 10.7 = 13.3$ mEq/L. But the patient's bicarbonate is 22 mEq/L, much higher than predicted. There is something "pushing the $[HCO_3^-]$ up." It is a metabolic alkalosis. Comparing the increase in the anion gap (16 mEq/L) to the decrease in the bicarbonate (only 2 mEq/L) allows us to identify a "hidden" metabolic alkalosis.

Example 5: A patient starts with AG 12, serum $[HCO_3^-]$ 24. ABG: pH 7.40, $[HCO_3^-]$ 24, and P_{CO2} 40. A ketoacidosis of 10 mEq/L develops. The new values are AG 22 mEq/L, $[HCO_3^-]$ 4 mEq/L, pH 7.08, and P_{CO2} 14 mm Hg. What is going on?
Answer:

Step 1: Bicarbonate is down, pH is down. There is a severe metabolic acidosis.

Step 2: What should the P_{CO2} be after compensation? Use the formula in **Fig. 9-1**: $(1.5 \times 4) + 8 = 14$. The fall in P_{CO2} is appropriate compensation for the metabolic acidosis. Therefore, there is no respiratory disorder present.

Step 3: The increase in the AG is 10 mEq/L. The decrease in the $[HCO_3^-]$ for an increase in AG of 10 mEq/L due to ketoanions would be expected to be roughly around 10 mEq/L, so we would expect the $[HCO_3^-]$ to be roughly $24 - 10 = 14$ mEq/L. But the $[HCO_3^-]$ is 4 mEq/L, which is 10 mEq/L *less* than predicted by the AG metabolic acidosis alone. Something is "pushing the $[HCO_3^-]$ down." What is it?

Answer: It is a second metabolic acidosis. *This second "hidden" acidosis is of the normal anion gap type.* This patient has *two* independent metabolic acidoses: a high anion gap acidosis and a normal anion gap metabolic acidosis.

Example 6: A patient presents to the emergency room in septic shock with the following: an anion gap that has increased from 12 to 30 (change in AG is 18) and a serum $[HCO_3^-]$ that has decreased from 26 to 4 (change in $[HCO_3^-]$ is 22). ABG: The P_{CO2} has fallen from 40 to 15, the $[HCO_3^-]$ has fallen to 4, and the pH has fallen from 7.40 to 7.05. What is your diagnosis?

Answer:

Step 1: The pH has fallen and the $[HCO_3^-]$ has fallen: metabolic acidosis.

Step 2: What should the P_{CO2} be? $P_{CO2} = (1.5 \times 4) + 8 = 14$. The measured P_{CO2} matches the predicted P_{CO2}. Therefore, there is no respiratory disorder present.

Step 3: A high anion gap acidosis in the setting of septic shock is most likely a lactic acidosis. The change in the AG is 18. We would expect the $[HCO_3^-]$ to fall by about $18/1.5 = 12$ mEq/L. This would lead to a $[HCO_3^-]$ of $26 - 12 = 14$ mEq/L. But the $[HCO_3^-]$ has fallen to 4. The $[HCO_3^-]$ is much less than predicted by a high anion gap acidosis alone. Something is pushing down the $[HCO_3^-]$ by an additional 10 mEq/L. The decrease in $[HCO_3^-]$ is explained by a coexisting normal anion gap metabolic acidosis.

These examples have pointed out how using the anion gap can identify additional "hidden" metabolic disorders in cases of lactic acidosis and ketoacidosis. In actuality, using the change in the anion gap to predict the change in bicarbonate is only an approximate method. Nevertheless, a significant deviation from this approximation suggests that an additional metabolic disorder may be present. That is:

- If the measured bicarbonate concentration is significantly higher than predicted by the increase in the AG, a "hidden" metabolic alkalosis may be present.
- If the measured bicarbonate concentration is significantly less than predicted by the increase in the AG, then a "hidden" normal anion gap metabolic acidosis may be present.

Exercises: Putting the Three Steps Together

For the sake of the following exercises assume that all the patients have the same baseline lab values: pH 7.40, P_{CO2} 40, $[HCO_3^-]$ 24, AG 12. All the changes and calculations for solving the cases should be based upon these baseline values.

Case 1

A patient presents with: pH 7.15, calculated $[HCO_3^-]$ 6 mEq/L, P_{CO2} 18 mm Hg, sodium 135 mEq/L, chloride 114 mEq/L, potassium 4.5 mEq/L, serum $[HCO_3^-]$ 6 mEq/L.

Step 1: This patient has a very severe metabolic acidosis.

Step 2: For a metabolic acidosis, what should the P_{CO2} be? We want to know whether this is a simple metabolic acidosis or if there is also a respiratory disorder present. The question we ask is: What should the P_{CO2} be after compensation? We answer this question with the formula for expected respiratory compensation for metabolic acidosis:

$$P_{CO2} = (1.5 \times 6) + 8$$
$$P_{CO2} = 9 + 8 = 17$$

The patient's P_{CO2} of 18 is close to the 17 we would expect for the appropriate respiratory compensation for a simple metabolic acidosis. Therefore, we conclude that there is no respiratory disorder present.

Step 3: The anion gap is AG = 135 − (6 + 114) = 15 mEq/L (normal). We are finished. There are no further steps if the AG does not suggest a high AG acidosis.

Answer: Simple normal anion gap metabolic acidosis. The differential diagnosis is listed in **Fig. 7-1.**

Case 2

A patient presents with: pH 7.08, $[HCO_3^-]$ 10, P_{CO2} 35, anion gap 14 mEq/L.

Step 1: The $[HCO_3^-]$ is 10, and the pH is 7.08. There is a severe metabolic acidosis.

Step 2: What should the P_{CO2} be? The P_{CO2} should be:

$$P_{CO2} = (1.5 \times 10) + 8 = 23 \text{ mm Hg}$$

The P_{CO2} of 35 mm Hg is much higher than we would expect! Therefore, there is something pushing the P_{CO2} up. It is a coexisting respiratory acidosis. There is a respiratory acidosis present as well as a metabolic acidosis.

Step 3: The anion gap is 14 (normal). We are finished.

Answer: Normal anion gap metabolic acidosis plus respiratory acidosis.

The patient's P_{CO2} is 35. This is much higher than predicted by the formula. Therefore, the patient has a respiratory acidosis which might represent "tiring out" of the patient's respiration and impairment of his ability to compensate for the metabolic acidosis. It could also be a clue to a coincident pulmonary process. *The rising P_{CO2} is a dangerous sign in metabolic acidosis, because further increase in the P_{CO2} could lead to a precipitous fall in pH.*

An important clinical note about the maximum compensation possible for a metabolic acidosis: In a young person, the maximum respiratory compensation (the lowest attainable P_{CO2}) is around 10–15 mm Hg. The value is about 20 mm Hg in an older person, indicating less ability to compensate by increasing ventilation. Therefore, there is a limit to the magnitude of respiratory compensation possible for a metabolic acidosis. A patient with a $[HCO_3^-]$ of 3 and maximal respiratory compensation will have a P_{CO2} of *roughly* $1.5 \times 3 + 8 = 12.5$ mm Hg. This is approximate because the compensation curve is not entirely linear at extremely low levels of $[HCO_3^-]$. The pH with this HCO_3^- concentration and P_{CO2} will be 7.00. To keep the P_{CO2} at 12.5 mm Hg takes a big effort. How long can the patient keep breathing deep and fast enough to hold the P_{CO2} at 12.5 mm Hg before tiring out? Suppose the patient begins to develop respiratory muscle fatigue, and the P_{CO2} creeps up to 20 mm Hg. The pH will plummet to 6.80!

The clinical point is that a young patient with severe metabolic acidosis and a P_{CO2} of 10–15 mm Hg or an older patient with a P_{CO2} of 20 mm Hg is "on the edge" of compensation; any further increase in P_{CO2} or further decrease in $[HCO_3^-]$ can mean disaster!

Case 3

A patient presents with: pH 7.49, $[HCO_3^-]$ 35, P_{CO2} 48, AG 16.

Step 1: The $[HCO_3^-]$ is increased and so is the pH: Metabolic alkalosis.

Step 2: What should the P_{CO2} be? We want to know if there is a respiratory disorder in addition to the metabolic alkalosis. Assuming a normal $[HCO_3^-]$ of 24 and a normal P_{CO2} of 40, the answer is: $P_{CO2} = 40 + .7 \times (35 - 24) = 47.7$. The patient's P_{CO2} is 48 mm Hg, which is what it should be for respiratory compensation for a simple metabolic alkalosis. Therefore, there is no coexisting respiratory disorder.

Step 3: The anion gap is 16 (normal).

Answer: Simple metabolic alkalosis.

Case 4

A patient presents with: pH 7.68, $[HCO_3^-]$ 40, P_{CO2} 35, AG 14.

Step 1: $[HCO_3^-]$ is up. pH is up. Metabolic alkalosis.

Step 2: What should the P_{CO2} be? The answer is: $P_{CO2} = 40 + .7 \times (40 - 24) = 51.2$ mm Hg. The patient's P_{CO2} is much less than predicted by the formula, even giving the P_{CO2} ± 5 mm Hg to account for variation in respiratory response to a metabolic alkalosis. Therefore, there is a coexisting respiratory alkalosis in addition to the metabolic alkalosis.

Step 3: The anion gap is 14 (normal). We are finished.

Answer: Metabolic alkalosis plus respiratory alkalosis.

There are *two* distinct acid-base disorders present, each with its own set of potential causes. The patient has a metabolic alkalosis secondary to one or more of the causes listed in **Fig. 8-1** *plus* a respiratory alkalosis. The causes for each disorder should be considered separately.

Case 5

A previously well patient presents with 30 minutes of respiratory distress and pH 7.26, P_{CO2} 60, $[HCO_3^-]$ 26, AG 14.

Step 1: The P_{CO2} is up. The pH is down. Respiratory acidosis. The history says acute.

Step 2: For respiratory disorders we ask: What should the $[HCO_3^-]$ be? Remember that the calculations for metabolic compensation are in terms of changes of 10 in P_{CO2}. The P_{CO2} is up by 20 which is 2×10. For an acute respiratory acidosis, the $[HCO_3^-]$ should change by 1 mEq/L for every 10 mm Hg increase in the P_{CO2}. Therefore, the $[HCO_3^-]$ should change by 2×1 mEq/L = 2 mEq/L. Using 24 as normal, the $[HCO_3^-]$ should become: $24 + 2 = 26$. Therefore, the compensation is appropriate, and there is no metabolic disorder.

Step 3: The AG is normal. We are finished.

Answer: Acute respiratory acidosis.

Case 6

Anxious. Can't seem to get enough air for last *4 days*. pH 7.42, P_{CO2} 30, $[HCO_3^-]$ 19, AG 16.

Step 1: P_{CO2} down. pH up. Respiratory alkalosis. The history indicates chronic respiratory alkalosis.

Step 2: What should the $[HCO_3^-]$ be? Remember that the calculations for metabolic compensation are in terms of changes of 10 in P_{CO2}. The P_{CO2} is down by 10 which is 1×10. For a chronic respiratory alkalosis, the $[HCO_3^-]$ should be down by 5 for every 10 mm Hg decrease in P_{CO2}. For this chronic respiratory alkalosis, the $[HCO_3^-]$ should be down by 1×5. The $[HCO_3^-]$ should be $24 - 5 = 24 - 5 = 19$. Therefore, the compensation is appropriate, and there is no metabolic disorder.

Step 3: The AG is normal. We are finished.

Answer: Chronic respiratory alkalosis. It is of interest that chronic respiratory alkalosis is the only simple disorder in which compensation can bring the pH back into the normal range (7.42 in this case).

Case 7

Short of breath. Two weeks. pH 7.38, P_{CO2} 70, $[HCO_3^-]$ 40, AG 16.

Step 1: P_{CO2} up. Respiratory acidosis. By history: chronic.

Step 2: What should the $[HCO_3^-]$ be? The P_{CO2} is up by 30 which is 3×10. For this chronic respiratory acidosis, the $[HCO_3^-]$ should increase by

$3 \times 3.5 = 10.5$. Using 24 as normal, the $[HCO_3^-]$ should become: $24 + 10.5 = 35.5$. In short: For this chronic respiratory acidosis the $[HCO_3^-]$ should be $24 + (3 \times 3.5) = 35.5$. The patient's $[HCO_3^-]$ is higher than it should be. Therefore, there is a modest metabolic alkalosis present as well. The high $[HCO_3^-]$ is not just compensation for the respiratory acidosis but is caused by a separate acid-base disorder: metabolic alkalosis.

Step 3: The AG is normal

Answer: Chronic respiratory acidosis plus metabolic alkalosis.

There are *two distinct* acid-base disorders present in this patient. Both disorders are pathologic — one is not *compensation* for the other, even though the pH may be close to normal. In other words, this patient has two processes going on at the same time that tend to *offset* each other: A metabolic alkalosis that is secondary to one of the causes listed in **Fig. 8-1** *plus* a respiratory acidosis. *Causes for each of the two disorders should be considered separately.*

Case 8

Try approaching case 7 the other way, starting with *metabolic alkalosis.*

Step 1: The $[HCO_3^-]$ is high: metabolic alkalosis.

Step 2: For a metabolic alkalosis, what should the P_{CO2} be? It should be $40 + .7 \times (40 - 24) = 51$. The P_{CO2} is much higher than this. Something is pushing it up: a respiratory acidosis. (Also, remember that for compensation for a metabolic alkalosis the P_{CO2} should not be higher than 55 mm Hg: It *is* higher than 55 mm Hg, indicating that a respiratory acidosis is present.)

Step 3: The AG is normal.

Answer: Metabolic alkalosis plus respiratory acidosis. The pH is often close to normal when *offsetting* disorders are present. Each disorder is pushing the pH in the opposite direction.

If given the choice of a metabolic or a respiratory disorder of equal severity being present at the same time, I will generally start my analysis of the data from the standpoint of the metabolic disorder first, because it will avoid the question of acute versus chronic and which formula to apply. The 3-step method will work either way, however, whether you begin with the metabolic disorder or the respiratory disorder. I just find that beginning with the metabolic disorder is sometimes less cumbersome.

Case 9

A patient presents with: pH 7.68, P_{CO2} 35, $[HCO_3^-]$ 40, AG 18.

Step 1: The pH is up and the $[HCO_3^-]$ is up: metabolic alkalosis.

Step 2: What should the P_{CO2} be? Apply the formula for metabolic alkalosis: $P_{CO2} = 40 + .7 \times ([HCO_3^-_{(measured)}] - [HCO_3^-_{(normal)}]) = 40 + .7 \times (40 - 24) = 40 + 11.2 = 51.2$. The patient's P_{CO2} of 35 mm Hg is significantly lower than predicted. Therefore, it is being pushed down by a respiratory alkalosis.

Step 3: The AG is 18. This AG is abnormal but it is less than 20 and we cannot make assertions about the presence of an AG acidosis. We are finished.

Answer: Metabolic alkalosis plus respiratory alkalosis. The pH is often severely abnormal when disorders are *synergistic*, each pushing the pH in the same direction.

Case 10

A patient presents with: pH 7.45, P_{CO2} 65, $[HCO_3^-]$ 44, AG 14. Short of breath for 3 days.

Step 1: Both the P_{CO2} and the $[HCO_3^-]$ are very high. The pH is normal. Let's call this a metabolic alkalosis because the pH is a little on the high side.

Step 2: What should the P_{CO2} be? For a metabolic alkalosis, the P_{CO2} should be $40 + .7 \times (44 - 24) = 54$. The patient's P_{CO2} of 65 is 11 mm Hg too high. Therefore: respiratory acidosis.

Step 3: The anion gap is normal.

Answer: Respiratory acidosis and metabolic alkalosis. Note that the pH is normal, while the P_{CO2} and the $[HCO_3^-]$ are both severely abnormal. This tells us immediately that there is a mixed disorder, because a patient cannot compensate all the way to a normal pH except in the case of a chronic respiratory alkalosis. This is an example of two disorders *offsetting* each other; that is, the disorders tend to cancel each other by pushing the pH in opposite directions. If you just eyeball the chemistries you might think that this patient has a simple respiratory acidosis with metabolic compensation: The data look as if there is only one disorder. Step 2 tells us that this is *not* just a simple respiratory acidosis with metabolic compensation. This patient has two distinct disorders.

Case 11

Same as Case 10, but start from the respiratory disorder: pH 7.45, P_{CO2} 65, $[HCO_3^-]$ 44, AG 14. Short of breath for 3 days.

Step 1: Both the P_{CO2} and the $[HCO_3^-]$ are abnormal. The pH is normal. Let's call this a respiratory acidosis—chronic because the history suggests that this has been going on for 3 days.

Step 2: What should the $[HCO_3^-]$ be? For a chronic respiratory acidosis, the HCO_3 should be $24 + (2.5 \times 3.5) = 24 + 8.75 = 32.75$. The $[HCO_3^-]$ of 44 mEq/L is too high. Therefore: metabolic alkalosis.

Step 3: The anion gap is normal.

Answer: Respiratory acidosis and metabolic alkalosis.

Case 12

A patient presents with: pH 7.65, P_{CO_2} 30, $[HCO_3^-]$ 32, AG 30. The patient has a temperature of 102 degrees and a blood pressure 80/50. He is diaphoretic. The urinalysis shows numerous white blood cells and many bacteria. A urine dipstick test for ketones is negative.

Step 1: pH is up. $[HCO_3^-]$ is up (metabolic alkalosis), and P_{CO_2} is down (respiratory alkalosis). Let's start with the metabolic alkalosis though it would work out either way.

Step 2: For metabolic alkalosis what should the P_{CO_2} be? $P_{CO_2} = 40 + .7 \times (32 - 24) = 45.6$. The patient's P_{CO_2} of 30 is much lower than 45.6. Therefore, respiratory alkalosis is also present.

Step 3: The anion gap is 30! Therefore, a high anion gap acidosis is present. We are up to three disorders. Captain, I don't think our engines can stand the heat! The most likely cause of high anion gap acidosis in this patient is lactic acidosis.

The change in the anion gap is $30 - 12 = 18$. We compare this to the change in $[HCO_3^-]$. The expected decrease in $[HCO_3^-]$ is roughly: Change in AG/1.5 = 18/1.5 = 12. The $[HCO_3^-]$ did not decrease, but is *up* by 8 mEq/L. It should be *down* by roughly 12 mEq/L. Therefore, the $[HCO_3^-]$ is about 20 mEq/L higher than we would expect. This means that there is a severe metabolic alkalosis acting to push the $[HCO_3^-]$ up by roughly 20 mEq/L in addition to a very severe high anion gap acidosis acting to push the $[HCO_3^-]$ down by roughly 12 mEq/L. The metabolic alkalosis and the metabolic acidosis tend to cancel each other, but they are both quite severe.

Answer: Metabolic alkalosis (severe), AG metabolic acidosis (severe), and respiratory alkalosis (moderate to severe). Note that the $[HCO_3^-]$ of 32 mEq/L does not look too bad at first glance. Calculating the AG and then comparing the increase in the AG to the decrease in $[HCO_3^-]$ was helpful in this case.

Case 13

A patient presents with diabetic ketoacidosis: pH 6.95, P_{CO_2} 28, $[HCO_3^-]$ 6, AG 32.

Step 1: The metabolic acidosis is so severe that this patient is in danger of cardiovascular collapse.

Step 2: What should the P_{CO_2} be? $P_{CO_2} = (1.5 \times 6) + 8 = 17$. The patient's P_{CO_2} is much higher than expected for this metabolic acidosis. The higher than expected P_{CO_2} indicates a respiratory acidosis, possibly secondary to respiratory muscle fatigue. Some might call this "inadequate compensation" instead of respiratory acidosis because the value of the P_{CO_2} is low, not high. They would be partially correct, but let's just stick to our original terminology so as not to gum things up. This is a severe metabolic acidosis in which the patient's respiratory compensation is beginning to "tire out."

Remember that patients cannot keep their P_{CO_2} in the 10–20 range indefinitely without eventually tiring out. Therefore, this patient has a metabolic acidosis and a respiratory acidosis secondary to respiratory muscle fatigue.

Step 3: The anion gap is 32. Therefore an anion gap acidosis is present. The increase in the anion gap is 20 mEq/L, supporting the diagnosis of AG acidosis. The predicted fall in the $[HCO_3^-]$ is roughly 20 mEq/L. The fall in the $[HCO_3^-]$ is 18, which is very close to 20. Therefore, the $[HCO_3^-]$ is close to what it should be for a ketoacidosis alone, and there is no "hidden" metabolic disorder.

Answer: AG acidosis secondary to diabetic ketoacidosis; respiratory acidosis due to respiratory muscle fatigue.

Case 14

A patient with recurrent episodes of small bowel obstruction presents with severe abdominal pain and vomiting: pH 7.33, P_{CO_2} 35, $[HCO_3^-]$ 18, AG 33. Urine dipstick negative for ketones. The blood pressure is 82/54 and the heart rate 116.

Step 1: $[HCO_3^-]$ down, pH down. Metabolic acidosis. Most likely a lactic acidosis. Looks pretty mild at first glance.

Step 2: What should the P_{CO_2} be? $(1.5 \times 18) + 8 = 35$. No respiratory disorder.

Step 3: The anion gap of 33 indicates that an anion gap acidosis is present. The increase in anion gap is 21. The decrease in the $[HCO_3^-]$ should be somewhere around $21/1.5 = 14$ for a lactic acidosis, but is only 6. Therefore, there is probably a "hidden" metabolic alkalosis acting to "push" the bicarbonate to a higher level.

Answer: Severe (18 mEq/L) anion gap metabolic acidosis plus metabolic alkalosis.

Case 15

A 21-year-old diabetic patient presents with vomiting and pH 7.75, P_{CO_2} 24, $[HCO_3^-]$ 32, AG 30. The urine is strongly positive for ketones and serum ketones are strongly positive.

Step 1: The pH is way up. $[HCO_3^-]$ is up. P_{CO_2} is down. Both of these are pushing the pH up. This is an example of a synergistic disorder in which the pH gets pushed the same way by both the P_{CO_2} and the $[HCO_3^-]$. This patient has life-threatening alkalemia. You could start with either the P_{CO_2} or the $[HCO_3^-]$ in this case. I prefer to start with the metabolic alkalosis.

Step 2: What should the P_{CO_2} be? $40 + .7 \times (32\text{-}24) = 45.6$ mm Hg. The patient's P_{CO_2} of 24 mm Hg is much lower than predicted. Severe respiratory alkalosis is present in addition to the metabolic alkalosis.

Step 3: The AG is 30. AG acidosis is present. The increase in the AG is 18. Accordingly, the $[HCO_3^-]$ should have fallen by roughly 18 mEq/L to the range of 6 mEq/L. But it is increased to 32!! The $[HCO_3^-]$ went up, not down! There is something pushing up the $[HCO_3^-]$ from the range of 6 mEq/L to 32 mEq/L!! Therefore: severe metabolic alkalosis. The initial eyeballing of the $[HCO_3^-]$ suggested that the metabolic alkalosis was "mild," but we can now see that it is very severe.

Answer: Respiratory alkalosis (severe), AG acidosis (severe), metabolic alkalosis (severe).

Case 16

This is a *totally* optional question: Reread the comment about maximum respiratory compensation for a metabolic acidosis that follows Case 2. How did I know that the pH of 7.00 in a patient with a $[HCO_3^-]$ of 3 and P_{CO2} 12.5 would plummet to 6.80 if the P_{CO2} increased to 20 mm Hg?

Answer: I used the Henderson-Hasselbalch equation

$$pH = 6.1 + \log ([HCO_3^-]/.03 \times P_{CO2})$$

and simply plugged in the values $[HCO_3^-] = 3$ and $P_{CO2} = 20$.

$$pH = 6.1 + \log (3/(.03 \times 20))$$

$$pH = 6.1 + \log (3 /.6) = 6.1 + \log (5) = 6.1 + .70 = 6.80$$

This equation is also useful to see if the pH, $[HCO_3^-]$, and P_{CO2} are consistent with each other or if there has been a lab error in measuring one of these variables. This formula is included because it is sometimes useful to have a way of verifying that the pH, $[HCO_3^-]$, and P_{CO2} results are correct, and to predict what would happen to the pH given a change in $[HCO_3^-]$ or P_{CO2}.

There are approximate methods available that don't involve using logarithms, but the Henderson-Hasselbalch equation is the easiest for me. I just bite the bullet and pull out my calculator. This formula is included because you might find it useful someday, but it is *not* important to working any of the exercises in this book.

CHAPTER 10. CASE EXAMPLES

Case 1

A 50-year-old 70 kg alcoholic man presents with 4 days of nausea, vomiting, and mild abdominal pain following a week-long drinking binge. He is unable to take anything by mouth. His mucous membranes are dry, and his vital signs reveal an orthostatic blood pressure drop with a rise in pulse. The following laboratory data are obtained: Na 134 mEq/L, K 3.1 mEq/L, [HCO_3^-] 20 mEq/L, Cl 80 mEq/L, glucose 86 mg/dl, BUN 52 mg/dl, Cr 1.4 mg/dl, amylase pending, serum ketones: high positive reading. ABG: pH 7.32, P_{CO2} 40 mm Hg, [HCO_3^-] 20 mEq/L. Urine sodium 7 mEq/L (low). Urine ketones: high reading. What is your diagnosis, and what do you do?

Answer: The history and laboratory studies suggest alcoholic ketoacidosis with hyponatremia secondary to volume depletion (vomiting) and hypokalemia secondary to vomiting and ketoacidosis. There may also be pancreatitis. There is a complex acid-base disorder, although the pH is only mildly depressed.

1. Complex acid-base disorder.

Step 1: pH is slightly decreased. [HCO_3^-] is slightly down: metabolic acidosis. P_{CO2} is "normal".

Step 2: For metabolic acidosis what should the P_{CO2} be? $P_{CO2} = (1.5 \times 20) + 8 = 38$. The measured P_{CO2} of 40 mm Hg is very close to this value, so no respiratory disorder is present.

Step 3: The anion gap is $134 - (20 + 80) = 34$! Therefore an anion gap acidosis is present. Now compare the *change* in the anion gap $(34 - 12 = 22)$ to the change in the [HCO_3^-] ($= 4$). The expected decrease in [HCO_3^-] based upon a ketoacidosis is in the range of 22 mEq/L. The [HCO_3^-] only decreased by 4 instead of 22 mEq/L. Therefore, there is a *metabolic alkalosis* acting to push the [HCO_3^-] up and a severe anion gap acidosis acting to push the [HCO_3^-] down. They tend to cancel each other, but they are both severe. The solution to the acid-base disorder is:

- Anion gap metabolic acidosis due to alcoholic ketoacidosis (review **Fig. 7-1** for other possibilities)

- Metabolic alkalosis due to vomiting (review **Fig. 8-1** for other possibilities)

I try to remember to *consider* ethylene glycol and methanol in an alcoholic patient with a high AG acidosis.

2. Hyponatremia. The patient has a history of vomiting and clinical evidence of ECFV depletion. The urine sodium is low. Review the causes of hyponatremia and the approach to the hyponatremic patient in **Figs. 3-1, 3-2, and 3-3.**

3. Hypokalemia. The hypokalemia is probably secondary to vomiting and ketoacidosis. A spot urine potassium to creatinine ratio > 20 mEq/gm would support urinary potassium loss (remember that hypokalemia is due to urine potassium loss in both vomiting and ketoacidosis). The serum potassium concentration of 3.1 mEq/L suggests a large deficit of as much as 400 mEq. The potassium concentration may fall with glucose administration, so potassium replacement should be started as soon as you know the patient is not anuric, and the potassium concentration rechecked in 2–3 hours. If the potassium concentration falls, then replacement should be increased (if the potassium concentration falls rapidly, then the glucose-containing saline solution could be held temporarily and 0.9% saline without glucose could be used if necessary). It is also important to measure a magnesium concentration in such a patient. Remember that potassium deficits cannot be replaced until the magnesium deficiency is corrected.

4. Orders: Patients *with alcoholic ketoacidosis* require glucose supplementation along with isotonic saline to reverse ketosis. Also, multivitamins, thiamine, and folate should be replaced in such a patient. IV glucose could precipitate an acute Wernicke's encephalopathy in this patient if thiamine (100 mg IM) is not given first. So, in an alcoholic, first give the thiamine; then start the fluids. The IV orders might look like:

100 mg thiamine IM stat and every day for three days
Liter #1: D5 0.9% saline with 30 mEq/L KCl 5 mg folate 1 Amp Multivitamins at 250 cc/hr.
Liter #2: D5 0.9% saline with 30 mEq/L KCl 5 mg folate 1 Amp Multivitamins at 175 cc/hr.
Liter #3: D5 0.9% saline with 30 mEq/L KCl at 175 cc/hr.

An IV order is not complete until the monitoring orders are written: Daily weight in the morning. Glucose, sodium, potassium, chloride, bicarbonate, blood urea nitrogen (BUN), and creatinine (Cr) *in 3 hours, 6 hours, 9 hours,* and in the morning. If the potassium concentration falls, then replacement should be increased.

Case 2

You are called to see a 40-year-old 60 kg woman who has had a generalized tonic-clonic seizure 36 hours after undergoing resection of a tubo-ovarian abscess. She is poorly arousable, but without focal neurological findings. She has the following laboratory data: Na 112 mEq/L, K 5.0 mEq/L, Cl 74 mEq/L, [HCO$_3^-$] 16 mEq/L, OSM$_{(meas)}$ 252 mOsm/L, pH 7.32, P$_{CO2}$ 32.

You check the preoperative lab results: Na 124 mEq/L, K 5.0 mEq/L, Cl 90 mEq/L, [HCO$_3^-$] 24 mEq/L. OSM$_{(meas)}$ 270 mOsm/L. What is your diagnosis and what would you do?

Answer:

1. Acute severly symptomatic hyponatremia with hypotonicity. There has been a large, rapid drop in the sodium concentration. You check what postop IV fluids the patient received: 6 liters of D5 0.45% saline over the last 36 hours. You stop the IV fluids immediately. This patient had significant hyponatremia on admission: Preoperatively, the sodium concentration was 124 mEq/L. Unexplained hyponatremia of this degree should be carefully evaluated preoperatively if possible. The evaluation does not take a long time in most cases, and should not delay the surgery needlessly. Review **Figs. 3-1, 3-2,** and **3-3.** The cause of the hyponatremia could be as simple as a thiazide diuretic or one of the medications or conditions listed in **Fig. 3-2.** The low preoperative serum sodium concentration and low measured osmolality indicate preexisting hyponatremia with hypotonicity. A patient with hyponatremia and hypotonicity should not receive hypotonic fluids under any circumstances. In general, hypotonic fluids should not be given to any patient postoperatively, either. Therefore, two serious errors contributed to this patient's cerebral edema. Determination of the underlying cause of the hyponatremia will have to wait for the time being, because we need to start emergency treatment.

Remember that the rate of fall of the serum sodium concentration is critical in determining whether there is severe brain edema and therefore whether there are symptoms or not. This patient has had a marked drop (12 mEq/L) in serum sodium over a period of only 36 hours, indicating that the patient's symptoms are due to cerebral edema secondary to acute hyponatremia. This patient may die if appropriate management is not begun immediately. If you cannot remember how to do the calculations for 3% saline infusion, you may temporarily begin with 3% saline at 50–100 ml/hr for a brief period until you can calculate a more precise rate. Carefully review the safety guidelines for rapid correction of acute, severely symptomatic hyponatremia with 3% saline given in Chapter 3.

Using the safety parameters and the values of serum sodium in this patient, calculate the amount of sodium to be given as 3% saline over 4 hours. At 4 hours what would you like the serum sodium to be? Looking at the safety guidelines for rapid correction of hyponatremia we would like the sodium to be approximately 116 mEq/L. Now use the equation:

$$Na \ (mEq) = ([Na^+_{(desired)}] - [Na^+_{(measured)}]) \times \text{Estimated Total Body Water}$$

$$Na^+(mEq) = (116 - 112) \times (.5 \times 60 \ kg)$$

$$Na^+ \ (mEq) = 120 \ mEq$$

So 120 mEq of sodium is to be given as 3% saline over the next 4 hours. Because 3% saline has 513 mEq sodium/L, the volume of 3% would be: $120/513 = .234$ L $= 234$ ml over 4 hours (about 60 ml/hr). The serum sodium concentration should be rechecked every 1–2 hours to monitor therapy.

2. Acid-base disorder.

Step 1: [HCO_3^-] down, pH down. Metabolic acidosis—most likely lactic acidosis.

Step 2: What should the P_{CO2} be? $1.5 \times 16 + 8 = 32$. No coexisting respiratory disorder.

Step 3: The anion gap of $112 - (12 + 74) = 22$ indicates that an anion gap acidosis is probably present. Comparing the anion gap with the previous day is very helpful here. The anion gap was 10 preoperatively. The increase of 12 in the anion gap indicates an AG acidosis. The expected decrease in the [HCO_3^-] is $12/1.5 = 8$, which exactly matches the decrease in our patient. Therefore, there is no "hidden" metabolic disorder. The AG acidosis is consistent with a lactic acidosis (urine ketones are negative).

Answer: Anion gap metabolic acidosis, probably a lactic acidosis due to the seizure. If this is the case, it should clear in 1–2 hours with no specific therapy.

Case 3

A 26-year-old diabetic man presents with polyuria, polydipsia, nausea and vomiting, following a bout with the "flu." The patient also states that he has been unable to hold anything down for the past 2 days. His temperature is 102 degrees, BP 118/74 and HR 100 (lying down), BP 90/60 and HR 120 (sitting with legs over the edge of the bed). The patient is in respiratory distress and is using his accessory muscles of respiration. You also note bilateral diffuse wheezing and rales at the right lung base. His laboratory studies: Na 122 mEq/L, K 4.5 mEq/L, HCO_3^- 15 mEq/L, Cl 80 mEq/L, glucose 325, BUN 30, Cr 1.2. pH 7.15, P_{CO2} 45, P_{O2} 68. CBC: Hgb./Hct. 12/36 WBC 15,000 UA: 1+ protein. Large ketones. Negative microscopic. Serum ketones: high positive reading.

What is your diagnosis and what would you do?

Answer:

1. Acid-base disorder.

Step 1: Metabolic acidosis (low [HCO_3^-], low pH).

Step 2: What should the P_{CO2} be? The P_{CO2} should be: $P_{CO2} = 1.5 \times 15 + 8 = 30.5$. The P_{CO2} is 45. This is much higher than expected, so there is a respiratory acidosis present. The respiratory acidosis renders the patient unable to adequately compensate for the metabolic acidosis. The respiratory acidosis is because of a coexistent pulmonary process (pneumonia/bronchitis?). Remember that failure of respiratory compensation is an ominous sign in a patient with metabolic acidosis.

Step 3: The anion gap is $122 - (15 + 80) = 27$ mEq/L. A high anion gap acidosis is present. Now, compare the increase in the anion gap $(27 - 12 = 15)$ to the decrease in $[HCO_3^-]$ $(24 - 15 = 9)$. The decrease in $[HCO_3^-]$ is 6 mEq/L less than what we would expect, but is within the "ballpark." The difference between the increase in the anion gap and the decrease in bicarbonate is not enough to really make an assertion that a metabolic alkalosis is present. Therefore, there are 2 acid-base disorders:

- Diabetic ketoacidosis (anion gap is 27) precipitated by the patient's respiratory infection.
- Respiratory acidosis from an as yet undiagnosed pulmonary process.

2. Asymptomatic hyponatremia.

You want to know the measured osmolality. It is 275. The measured osmolality confirms that *hyponatremia with hypotonicity* is present. Just because the glucose is high does not tell you that you have hyponatremia with hypertonicity. The glucose is 325 mg/dl. The "corrected" sodium concentration after correction for the elevated glucose would be only: $122 + (1.6 \times 2) = 125$. This is not hyponatremia with hypertonicity. Most likely, the hyponatremia is from ECFV depletion from protracted vomiting with continued water ingestion, although other potential causes should be considered. Because the hyponatremia is asymptomatic, we do not need to aggressively raise the serum sodium. In fact, rapid correction of this patient's hyponatremia could lead to the ODS. Clinically, this patient seems to have a chronic hyponatremia, which has developed over the past several days. Carefully review **Fig. 3-1, Fig. 3-2,** and **Fig. 3-3.** In addition to ECFV depletion, possible causes of hyponatremia include impaired GFR (Cr is 1.2 mg/dl, which rules this out), thiazide diuretics (no history of this), or SIADH from his pulmonary process (remember that this patient may have pneumonia). A number of medications can cause SIADH. The history does not reveal that he is taking chlorpropamide, an oral hypoglycemic agent that can produce SIADH. The suspicion of ECFV depletion could be further substantiated by obtaining a spot urine sodium. It will likely be <10 mEq/L. Sometimes, the diagnosis of hyponatremia is unclear: It is not certain whether the patient has mild ECFV depletion or SIADH. The response of the sodium concentration to administration of 0.9% saline with close monitoring of the sodium concentration and ECFV status may be helpful diagnostically. In ECFV depletion, the sodium will often begin to correct rapidly. In SIADH, the sodium concentration will usually not change much. Because of the possibility of ECFV overload, 0.9% saline administration is not recommended as a *routine* part of the diagnosis of hyponatremia, but may be helpful in this patient.

3. Potassium depletion.

It is likely that the potassium of 4.5 mEq/L is masking potassium depletion in a patient with diabetic ketoacidosis and a pH of 7.15 who has been vomiting for the past several days. Potassium supplementation is necessary as soon as it is established that the patient is not anuric.

4. Management considerations.
The pH is 7.15 and the patient shows signs of respiratory compromise and an inability to appropriately compensate for the metabolic acidosis. If such a patient should slow his respirations acutely, his pH would plummet and he could sustain a cardiac arrest. Under these conditions, I would *consider* HCO_3^- therapy, although there are reasons against giving HCO_3^- immediately in this case:

- The ketones will be converted to HCO_3^- by the liver once ketosis is reversed by insulin and IV fluids.
- Rebound metabolic alkalosis can occur.
- HCO_3^- can acutely raise the pH, causing potassium to shift into cells. If the potassium concentration were lower in this case, say around 3.5, this would be an even more important consideration.

The patient should be treated with IV insulin, KCl, and 0.9% saline. Once it is established that he is not anuric, the initial fluid orders should look something like: IV: 0.9% saline with 30–40 mEq/L KCl at 250 ml/hr. The electrolytes and ABG should be rechecked in 2 hours. An increased rate of potassium replacement may be required, depending upon the repeat potassium concentration. The serum calcium, magnesium, and inorganic phosphate concentrations should also be checked in this patient and multivitamins and folate added to the first liter of fluids. Thiamine should also be given. Diagnosis and appropriate treatment of the pulmonary process causing the respiratory acidosis and continuous monitoring of respiratory status and pH will be key in this case. Should serious respiratory decompensation occur, the patient could be intubated and mechanically ventilated.

Case 4

A 75-year-old woman is referred because of back pain. Her laboratory studies: Na 124 mEq/L, K 4.2 mEq/L, Cl 100 mEq/L, HCO_3^- 24 mEq/L, BUN 28, glucose 90 mg/dl. What do you think?
Answer: You calculated the anion gap, right? Of course you did. The low anion gap (zero) is a clue to the possibility of multiple myeloma. In some patients, the paraproteins are positively charged and increase the unmeasured cations (UC) so that the anion gap decreases according to the relation $AG = UA - UC$. The protein concentration in this patient is 12 gm/dl, and the hyponatremia is the result of pseudohyponatremia secondary to multiple myeloma paraproteinemia. The lab was not using a sodium electrode for some reason. Serum osmolality is normal. The patient is not hypotonic: a measured osmolality is 285 mOsm/L. Patients with pseudohyponatremia have an increased osmolal gap. The osmolal gap is:

$$285 - (248 + 90/18 + 28/2.8) = 22 \text{ mOsm/L}$$

I added this exercise mainly because you may see this woman again on a board exam. I also needed to remind myself to calculate the AG on *every* set of electrolytes.

Case 5

A 65-year-old woman presents with mild forgetfulness. She lives with her husband Fred and their cats Sidney and Tabbert and has not been to a doctor's office for 25 years. Laboratory studies: Na 124 mEq/L, K 4.2 mEq/L, Cl 90 mEq/L, HCO_3^- 24 mEq/L, Cr 1.0 mg/dl, BUN 14 mg/dl, glucose 90 mg/dl. What do you think?

Answer: Review **Figs. 3-1**, **3-2**, and **3-3**. You would like to know the measured osmolality. It is 260 mOsm/L, and the osmolal gap is 2 mOsm/L. This patient has hyponatremia with hypotonicity. Hyponatremia with hypotonicity is always due to impaired water excretion in the presence of continued water intake. Systematically:

1. Is there renal failure? The creatinine is normal, which rules out renal failure.
2. Is there evidence of abnormally increased or decreased ECFV? We look carefully for an edematous state or for evidence of ECFV depletion. There is none. We measure the urine sodium concentration. It is 45 mEq/L, which is against ECFV depletion or an edematous disorder.
3. Is the patient taking thiazides? In an elderly woman, hyponatremia may result from thiazide diuretics given to treat hypertension. We do not have a history of this.
4. Is there evidence of a disorder or is the patient taking a medication capable of causing SIADH (carefully review **Fig. 3-2**)?
5. Is there evidence of adrenal failure or hypothyroidism? When in doubt, order the appropriate assays.
6. Finally, we consider the so-called "tea and toast" diet.

If the solute excretion is low and the water intake high enough in a patient with impaired urinary dilution, hyponatremia may develop. In adult Americans, the average daily obligatory solute load is around 600–900 mOsm and consists mainly of urea and electrolytes (mostly sodium and potassium). The normal kidney is able to dilute the urine to as little as 50 mOsm/L or to concentrate the urine to as high as 1200 mOsm/L. Therefore, the urine volume in 24 hours could be as high as approximately 900 mosm/50 mOsm/L = 18 L in a maximally dilute urine (state of water excess) and as low as 600 mosm/L/1200 mOsm = .5 L in a highly concentrated urine (state of water conservation).

An elderly patient ingesting a diet poor in protein and NaCl may have impaired water excretion resulting from a decreased solute excretion. The solute load might be around 300 mOsm/day. Let's say that the minimum urine concentration that this elderly patient can attain is 150 mOsm/L, instead of

50 mOsm/L. How much water can this person drink without developing hyponatremia? The *rough* approximation is: 300 mosm/150 mOsm/L + 1/2 L (insensible) = 2 1/2 L! Consequently, our elderly woman with impaired diluting capacity and poor solute intake might develop hyponatremia if drinking more than about 2 1/2 L/day! The imbalance created by a low solute load, impaired ability to produce a dilute urine, and increased water intake is the mechanism behind the hyponatremia of the "tea and toast" diet. If the solute intake were increased from 300 to 600 mOsm/day, then the water intake could be increased to about 600/150 + 1/2 L = 4 1/2 L/day, and the patient would be much less likely to develop hyponatremia.

Case 6

A 45-year-old woman with diabetes mellitus is referred because of hyperkalemia. She feels well. There is no history of weakness. Her medications include captopril 25 mg TID, ibuprofen 400 mg TID PRN, glyburide 10 mg every day, and a multivitamin. She states that recently she has begun training for a triathalon. Her laboratory studies: Na 138 mEq/L, K 6.3 mEq/L, HCO_3^- 20 mEq/L, Cl 100 mEq/L, BUN 35 mEq/L, creatinine 2.1 mg/dl, glucose 160 mg/dl. UA 1 + protein, 2 + glucose, Sediment: Negative.
What is your approach to the differential diagnosis and what do you do?

Answer:
1. Stop all administration of potassium.
2. Obtain a stat ECG.
3. Quickly make a mental inventory of possible "hidden" sources of potassium and potential causes of hyperkalemia such as:
   Potassium penicillin
   Salt substitutes (many contain KCl)
   Hemolysis
   Transfusion
   Gastrointestinal hemorrhage
   Rhabdomyolysis
   Burns
   Major surgery
   Medications that can cause hyperkalemia
4. Send repeat potassium (drawn without tourniquet to reduce the chances of hemolysis).
5. Review all the medications taken by the patient
6. Determine the underlying cause of the hyperkalemia.

The causes of hyperkalemia are reviewed in **Fig. 6-1.** This patient most likely has syndrome of hyporeninemic hypoaldosteronism (SHH) that is aggravated by the ibuprofen and captopril. Primary adrenal failure and tubular unresponsiveness to aldosterone should also be considered. The creatinine concentration of 2.1 mg/dl indicates that some degree of renal failure is also present, which can contribute to the potassium excretory deficit. Remember that the causes of hyperkalemia may be additive, and that a patient may have more than one basis for hyperkalemia. The acute management of this patient is dictated by whether or not there are significant ECG changes and by the severity of the hyperkalemia. The chronic management begins with dietary counseling and a low-potassium diet, because the primary problem is renal retention of potassium. The hyperkalemia of SHH will generally respond to loop diuretics or to the combined use of loop diuretics and the potent mineralocorticoid fludrocortisone. The main side effect of fludrocortisone is renal sodium retention and volume overload. Therefore, this medication should be started under close observation with attention to body weight and observation for signs of ECFV excess. In general, fludrocortisone should be avoided in patients with significant history of congestive heart failure or other conditions associated with sodium retention. Drugs known to produce hyperkalemia should be stopped.

Case 7

A 50-year-old woman was admitted to the hospital with protracted nausea, vomiting, and abdominal pain. Abdominal X-rays revealed an ileus, which resolved with nasogastric suction and IV fluids (0.9% saline with 30 mEq/L KCl). She says that her abdominal pain, which had initially improved with nasogastric suction and IV fluids, has now returned. She now has a temperature of 101.6 and her blood pressure has fallen from 130/86 to 86/52. The abdomen is very tender, and no bowel sounds are present. Her laboratory studies: Na 140 mEq/L, K 4.5 mEq/L, Cl 80 mEq/L, HCO_3^- 25 mEq/L, pH 7.40, P_{O2} 100, P_{CO2} 40, HCO_3^- 25 mEq/L. What is your diagnosis?

Complex acid-base disorder.

Step 1: On inspection of the laboratory studies, there is no *obvious* acid-base disorder present. The pH, P_{CO2}, and $[HCO_3^-]$ are all normal.

Step 2: Because there is no apparent acid-base disorder present, appropriateness of compensation is not an issue.

Step 3: The anion gap is $140 - (25 + 80) = 35$! Therefore, a severe (most likely lactic) anion gap metabolic acidosis is present. This acidosis is probably the result of bowel ischemia. Why is the $[HCO_3^-]$ normal? Because there is an equally profound metabolic alkalosis present, which is "masking" the metabolic acidosis. We calculate the change in anion gap and compare it to the change in the $[HCO_3^-]$. The change in anion gap is $35 - 12 = 23$. Therefore, in this lactic acidosis, the $[HCO_3^-]$ should be around $25 - 23/1.5 = 25 - 15.3 = 9.7$! There is an opposing metabolic alkalosis pushing up the $[HCO_3^-]$ by around 15.3 in this case, and therefore the normal $[HCO_3^-]$ disguises two severe acid-base disorders:

- Anion gap acidosis from ischemic bowel.
- Metabolic alkalosis from vomiting and nasogastric suction.

It is important to follow the 3 steps for *every single* set of acid-base chemistries that you evaluate, even if everything looks normal at first glance. Calculating the anion gap was central to solving this case.

INDEX

NOTES

NOTES

NOTES

NOTES

RAPID LEARNING AND RETENTION THROUGH THE MEDMASTER SERIES:

CLINICAL NEUROANATOMY MADE RIDICULOUSLY SIMPLE, by S. Goldberg
CLINICAL BIOCHEMISTRY MADE RIDICULOUSLY SIMPLE, by S. Goldberg
CLINICAL ANATOMY MADE RIDICULOUSLY SIMPLE, by S. Goldberg
CLINICAL PHYSIOLOGY MADE RIDICULOUSLY SIMPLE, by S. Goldberg
CLINICAL MICROBIOLOGY MADE RIDICULOUSLY SIMPLE, by M. Gladwin and B. Trattler
CLINICAL PHARMACOLOGY MADE RIDICULOUSLY SIMPLE, by J.M. Olson **OPHTHALMOLOGY MADE RIDICULOUSLY SIMPLE**, by S. Goldberg
PSYCHIATRY MADE RIDICULOUSLY SIMPLE, by W.V. Good and J. Nelson
CLINICAL PSYCHOPHARMACOLOGY MADE RIDICULOUSLY SIMPLE, by J. Preston and J. Johnson
ACUTE RENAL INSUFFICIENCY MADE RIDICULOUSLY SIMPLE, by C. Rotellar
USMLE STEP 1 MADE RIDICULOUSLY SIMPLE, by A. Carl
USMLE STEP 2 MADE RIDICULOUSLY SIMPLE, by A. Carl
USMLE STEP 3 MADE RIDICULOUSLY SIMPLE, by A. Carl
BEHAVIORAL MEDICINE MADE RIDICULOUSLY SIMPLE, by F. Seitz and J. Carr
ACID-BASE, FLUIDS, AND ELECTROLYTES MADE RIDICULOUSLY SIMPLE, by R. Preston
THE FOUR-MINUTE NEUROLOGIC EXAM, by S. Goldberg
MEDICAL SPANISH MADE RIDICULOUSLY SIMPLE, by T. Espinoza-Abrams
THE DIFFICULT PATIENT, by E. Sohr
CLINICAL ANATOMY AND PHYSIOLOGY FOR THE ANGRY HEALTH PROFESSIONAL, by J.V. Stewart
CONSCIOUSNESS: HOW THE MIND ARISES FROM THE BRAIN, by S. Goldberg
PREPARING FOR MEDICAL PRACTICE MADE RIDICULOUSLY SIMPLE, by D.M. Lichtstein
MED'TOONS (260 humorous medical cartoons by the author) by S. Goldberg
CLINICAL RADIOLOGY MADE RIDICULOUSLY SIMPLE, by H. Ouellette
NCLEX-RN MADE RIDICULOUSLY SIMPLE, by A. Carl
THE PRACTITIONER'S POCKET PAL: ULTRA RAPID MEDICAL REFERENCE, by J. Hancock
ORGANIC CHEMISTRY MADE RIDICULOUSLY SIMPLE, by G.A. Davis
CLINICAL CARDIOLOGY MADE RIDICULOUSLY SIMPLE, by M.A. Chizner
PSYCHIATRY ROUNDS: PRACTICAL SOLUTIONS TO CLINICAL CHALLENGES, by N.A. Vaidya and M.A. Taylor.
MEDMASTER'S MEDSEARCHER, by S. Goldberg
PATHOLOGY MADE RIDICULOUSLY SIMPLE, by A. Zaher
CLINICAL PATHOPHYSIOLOGY MADE RIDICULOUSLY SIMPLE, by A. Berkowitz

Try your bookstore. For further information and ordering send for the MedMaster catalog at MedMaster, P.O. Box 640028, Miami FL 33164. Or see http://www.medmaster.net for current information. Email: mmbks@aol.com.